I0004101

Data Driven
System Engineering

Automotive ECU Development

James Wen

Copyright © 2021 James Wen
All rights reserved.
ISBN: 979-8-9856-2490-8

DEDICATION

For my 40 years career history.

ACKNOWLEDGMENTS

Great appreciation for my lovely wife's understanding and support, and this book acts as the milestone of knowing each other for 40 years.

1 Preface

1.1 Contents

What is the "Data Driven System Engineering" about?

Every computing system has two, and only two attributes: Data Value and Data timing, which fully represent the system functionalities from the system external behavior point of view. And the system development goal is to realize those attributes for each required output data.

The data driven system engineering is the approach to develop the system by focusing on the two attributes mentioned above, in which, the data values are derived by the system operation concept design, and the data timing is derived by the system latency design, so that the system development activities including the requirement engineering, system structure design, functional reliability design, functional availability design, system verification can be optimized significantly.

Figure 1.1-1 Computing System

For every computing system, it can be described using three elements illustrated in Figure 1.1-1 Computing System above, which consist of:

- input
- output
- process

The process in Figure 1.1-1 Computing System above comprises one or more calculations that transfer the input via some middle results to become the output by using designed operations, such as mathematic operations, logic operations, fuzz operations, AI operations, then the relationship between the output and input can be described as the calculations below:

Output = f(Input, Middle Result 1, … Middle Result n)

And in the computing system, the input signal, output signal and middle results can be represented by the data:

- Input Data, representing the input signals
- Middle Data, representing the middle results
- Output Data, representing the system final result

The output data depends totally on the input data, the middle data and the operations in the calculations, and every input data, middle data and the operation in the calculations has effect to the output data, and anything else that is not in the relationship will not have any effect to the output data.

And from the system functionality point of view:

- the output data represent the required system external behaviors, which are the system functionalities under the input data.
- the middle data present the middle functionalities during the transferring and transforming from the input data to the output data.

- the operations' effects are to transfer and transform data from one location or form to another location or form.

From the correctness point of view, if the data derived from the operations are correct, then the operations must be correct. So, to the development activities that only concern the functionality correctness, the attention can be only paid to the derived data, i.e., the system engineering activities will be most efficient if they apply only on the elements that have the effects to the output data.

Based on the approach above and the mechanisms in the AUTOSAR, this book provides a full range of system engineering development activities:

- Requirement Elicitation
- Requirement Engineering
- System Architecture Design
  - System Operation Concept Design.
  - System Structure Design
  - Electronic Architect Design
  - Functionality Allocation
  - Failure Mode and Effect Analysis (FMEA)
  - Safety
  - Cybersecurity
- System Verification
  - System Integration and Verification
  - System Black Box Verification

each of which has its own clearly defined scope and approach, which is different from the conventional development, in some cases even different from some ISO standards, for example:

- About the safety development, in this book, the safety requirements for every part in a vehicle are cascaded from the vehicle safety requirements, which is different from the Concept Phase in the Part 3 of ISO 26262.
- About the error detection, in this book, there are only two types of errors to be detected in a computing system: Data Value error and Data Timing error.

There are the corresponding check lists for each section above, which can be taken as either reading summaries or implementation guidelines.

The following sections are the brief descriptions.

1.1.1 Requirement Elicitation

The main system requirements are defined as they are the system descriptions from the black box point of view, they should describe the external behaviors of the system under development, everything inside of the ECU under development belongs to the ECU design, which are the ECU designers' decisions. This book categorizes the requirements as:

- Technical Requirements
  - Computable
  - Non-computable
- Non-Technical Requirements

And this book introduces the "Two Steps" requirement elicitation approach:
- the first step covers the "Width", which can be done easily and quickly to cover all the signals that the ECU needs to handle and provide the complete coverage of the ECU capability.
- the second step goes into the "Depth" which can provide the detailed and categorized functionalities below according to the modulization of 3.4.2 System Structure Design based on the input and output signals from the first step.
  - Feature function
  - Maintenance function represented by the diagnostic services
  - Input and output interfaces and signals
  - Safety
  - Cybersecurity
  - Quality Management

1.1.2   Requirement Engineering

This book makes the goal and scope of requirement engineering in the computing system development specific, accurate and measurable:
- Scope: the requirement engineering is to use the computer executable information to describe the system under development illustrated in Figure 1.1-1 Computing System, which consists only two types of information: Signal and Test Case.
- Measurement:
  - Signals, either input or output signals, shall be computer readable.
  - Test cases shall be executable in the system under development.

The benefit of making the requirements measurable is to have the defined way to measure the requirements' quality.

If there is anything that is not measurable, then it cannot be trustable.

The book emphasizes that the "requirement engineering" is an engineering activity, and for every engineering activity, there must be two aspects that should be done:
- Analysis
- Design

For every requirement item in the requirement specification, it must have some relationship(s) with other items, i.e., there is not isolated requirement in any specification. So, the requirement engineering should figure out the relationships between them, especially the relationships between the output signals with others, then optimize or design the better way to realize the required output signals.

1.1.3   System Architecture Design

The goal of system architecture design is to provide the platform that transfers and transforms the input signal to become the required output signal via some middle data.

This book introduces the following system functional modulizations based on the AUTOSAR that satisfies a generic automotive ECU structure:
- Feature Function
- Diagnostic Service
- Cybersecurity Function
- Serial Signal Manager

- Application Mode Manager
- AUTOSAR

In this book, the system architecture design is based on the data flow which is defined by the relationships between the input data and output data illustrated in Figure 1.1-2 Computing System with Multiple Signals below, the benefits of which are:
- All development activities share the same concept, which results in high development efficiency.
- The granularity of development and specifications is defined by the concept, which results in persistent and full system coverage without any redundancy.
- The correctness of development, such as reliability, safety, will apply only on the data that represent fully the system functionalities, which results in another level high efficiency.
- The solutions are clearly and completely defined approach with recursive mechanisms to develop the system.

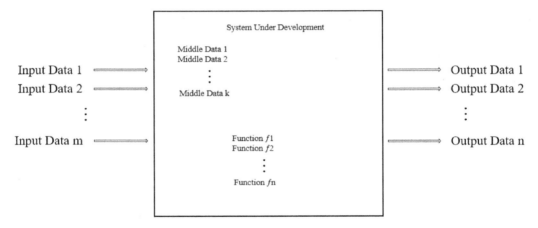

Figure 1.1-2 Computing System with Multiple Signals

so, the relationships between those signals in Figure 1.1-2 Computing System with Multiple Signals above can be described as:

Output Signal 1 = $f1$ (some Input Signal, some Middle Result);
Output Signal 2 = $f2$ (some Input Signal, some Middle Result);

$\bullet \ \bullet \ \bullet$

Output Signal n = fn (some Input Signal, some Middle Result);

The concept represented by the relationships above is the System Operation Concept, which is mandatory for every computing system development because the output data are derived from them. Based on which, the system architecture design activities in this book, such as functional definitions, functionality allocations, safety development, will focus only on the elements in the concept, and most of them will focus only on the data in the system operation concept, which makes the development efficient.

The commonly used system architecture design is based on the developers' experience, and the specifications are specified either using the text tools, such as IBM DOORS or PTC Integrity, or the notation tools, such as the SysML that includes 9 types of diagram, the issues of which are that there is neither the clearly defined explicit and complete approach to design the system architecture and specifications, nor is there the clearly defined explicit and complete method to fully cover all the system, which will cause issues in the next development.

For the text specified specifications, the issues will include that the text specifications are prone to ambiguous and incomplete, and it is difficult to figure out the logic relationships in the specifications, whose consequence is that the specifications may be inconsistent, incomplete and inaccurate, then it will cause the issues in the following development.

For the notation specified specifications, the issues include that it is difficult to fully specify the system functionalities, and it is difficult to use the notations in the entire development team, and it is difficult to figure out the relationship in all the diagrams used in the development.

1.1.4 Failure Mode and Effect Analysis

In this book, the FMEA will make use the system operation concept to find the failure modes, failure causes, effects analysis, and risks analysis. Taking the relationship below as an example:

Output Signal 1 = f(some Input Signal, some Middle Result);

by which, the relationship between the output signal 1, input signal and middle result is defined by the operation of f, then the failure modes, failure causes and failure effects of the data will impact each other according to f, in this way, FMEA development logic is exactly the same as the system operation concept as it is supposed to be, so that it can be accurate and efficient.

Another improvement in the FMEA is the risk analysis consisting of assigning a severity level, a probability level and a controllability level for each failure mode in the system under development, in this book, the risk analysis is carried over from ISO 26262, in another words, the classifications of severity, probability and controllability are exact same as the ones in ISO 2626, and those activities are mandatory in the automotive safety system development, and they are clearly defined and standardized in the standards, so it will be efficient to re-use those concepts.

The FMEA method above can be done recursively to any data that need to be decomposed further into decompositions as the development progresses.

The benefits of the FMEA driven by the data consist of making use the definitions from the system operation concept, and all the FMEA activities apply only on the data, and the process of doing the FMEA above is clearly and completely defined and optimized, the result of which will be efficient, accurate, complete and consistent.

1.1.5 Safety

The safety including Safety of The Intended Function (SOTIF) development in a computing system can be fully covered by the following three aspects:
- reliability
- functional availability
- product quality

which are important for all products, even for the non-safety systems.

The safety mechanisms are to ensure that the system operates within the system performance limitations and prevent the system from internal errors, and the internal errors may occur anywhere in the system under development, so the safety development is involved in every step in the development.

Although there are some quantitative hardware criteria from ISO 26262 about Single Point Fault Metric (SPFM), Latent Fault Metric (LFM) and the Probabilistic Metrics for Hardware Failures (PMHF) listed in the Table 3.4-15 Hardware Fault Metrics, however, the criteria to measure if a system is safe are very vague:

- The system safety measurements, especially the software safety measurements are not specified.
- Only very high-level activities are specified in ISO 21448, ISO 26262 Part 4: Product development at the system level and the ISO 2626 Pat 6: Product development at the software level, but those activities are highly dependable to interpretation and implementation, which is very difficult to make accurate adjudgment.

To improve it, this book provides solutions from the Data Driven point of view:

- Full system error detection: provide the solution to fully detect the errors in the system under development, which consist of only two types of errors: value error and timing error.
- Approach to achieve the safety: provide the solution to achieve the three aspects that are needed by the safety including Safety of The Intended Function (SOTIF) in the system under development.
- ISO 21448 and ISO 2626 compliance: provide the rationales for the solutions in the book to meet the ISO 21448 and ISO 26262 requirements, which can be a leading example for similar projects to achieve the compliance.

1.1.6 Cybersecurity

This book provides the approach to achieve the cybersecurity in the automotive Electronic Control Units (ECUs) that are inside of the vehicle network, consisting of:

- Trusted contents in the ECU
- Authenticated access to the ECU
- Authenticated communication with the ECU

Those implementations are reasonable enough to satisfy the UN ECE 155 / 156 which are mandatory for the vehicles to be sold in European, North American, Japanese and South Korean markets.

The implementations in the book optimizes the cybersecurity threat impact analysis mechanisms in the ISO 21434 by removing the operational impact that, from the book point of view, can be covered either by the financial impact or by the safety impact.

The scope of cybersecurity development in the book is only limited to the automotive ECUs that are allocated in the vehicle network behand the Gateway ECU illustrated below.

The network characteristics in a vehicle illustrated in Figure 2.2-1 Vehicle ECU Network leads to the significant different security & privacy requirements to the automotive ECUs comparing with the daily used computers that are connect to the internet and surf the webs all the time, among which, the main differences are:

- Execution Contents: all contents in the automotive ECUs are extremely controlled including the updating and modifying, however, for the daily used computers, the updates or download contents from various websites happen time to time.
- Communication: The communications between the automotive ECUs inside of the vehicle network behand the Gateway ECU illustrated in Figure 2.2-1 Vehicle ECU Network are pre-defined, such as which ECUs should send what messages to which ECUs. However, for the daily used computers, they may go to any websites on the internet to interact with any software.

So, in a vehicle network, only the Gateway ECU and On-Board Diagnostic (OBD) Interface ECU are similar as the daily used computers from outside of vehicle point of view, those ECUs need to implement the cybersecurity protections about the threats, vulnerabilities and attacks as the daily used computers do, which are out of this books' scope. All other ECUs that are behand the Gateway ECU are allocated in the strictly managed local vehicle network, which are the focus in this book.

1.1.7   Verification

The book clearly defines the scope, the purposes and the approaches of system verifications.

In the system development, the verification consists of two parts:
- System Black Box Verification, which is the verification to the test the system from black box point of view, which consists of only types of tests:
    - Environment persistence test
    - Time persistence test
- System Integration Verification, which is to test all the system functionalities.

That is, the system integration verification is the system functional test, the system black box verification is the system performance test, because:
- By the system integration test, not only the system external behaviors but also the system internal behaviors are disclosed, so that the result correctness can be accurately adjudged, which cannot be done by the system black box test that discloses only the external behaviors.
- By the system black box test, since the test environment is the real product operation environment, so the performance result is accurate.

Among the verification, the system verification is a time-consuming activity, the system integration verification is a technical complicated activity.

The system integration is prone to issues, such as:
- The integrated component interfaces don't match.
- The required operation environment doesn't match including the partitions, cores.
- The required time conflicts each other, either the running sequences don't match, or the scheduled time is not correct

The issues are inherited from the weaknesses in the system architecture design development using the conventional ways, which are analyzed in the section of 3.4 System Architecture Design.

In this book, the system integration process is defined by the system operation concept, in which, each data is an integration node, each formula is an integration path, so that the components' interfaces, the integration environment including time, location are clearly and completely defined.

1.1.8   Programming language-Oriented Diagram Notation

This book introduces the simple and efficient programming language-oriented diagram notation.

The purposes of Programming language-oriented diagram notation are:
- To make the design close to the required software program as much as possible
- To make the notation easy to use as much as possible

The Programming language-oriented diagram notation is used to design the system under development in diagrams that is simple and straightforward notations consisting:
- Executable Procedure Node (EPN)
- Executable Procedure Node Name and I/O Parameter
- Data definition
- Action
- Transitions between the EPNs

The EPN structure is illustrated as in Figure 1.1-3 Executable Procedure Node Structure below, which is totally the same as a function definition in any programming language.

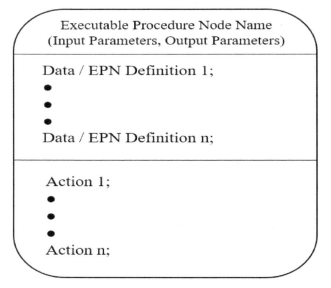

Figure 1.1-3 Executable Procedure Node Structure

The syntax for the notation is the syntax of the programming language selected for the project, in another words, the system developers can choose any syntax to define the notion contents. If the system under development is using C, then the EPN, data definition, the action instructions are defined using the select C programming language; if the C++ is selected, then the syntax is the C++; if java is selected, then the syntax is java, and so on.

The system under development can be described using the EPN flow illustrated in Figure 1.1-4 EPN Flow below, which is very similar as the State Machine diagram in the

SysML except the syntax.

So that, the design using this notation will be very close to the final product, the source code for the system under development will simply be the collections of all the EPNs if all the EPNs are described in enough details using the selected syntax.

The intention of the notation is to simplify the system design notation, and comparing with the nine diagrams in the SysML, the notation here is enough to do the system design, even in the detailed software design, it will still be good under the help of timing diagram.

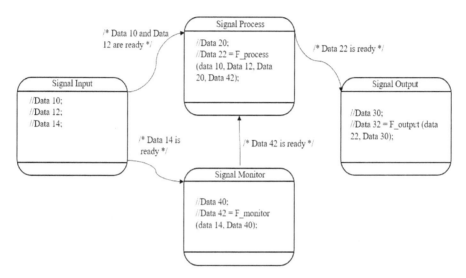

Figure 1.1-4 EPN Flow

1.1 Contents

1.2 Intention

There are the following intentions for the author to write this book:

1.2.1 Improve the System Development Process

System Engineering is the slow turtle in the development race!

Among the three domains in the computer engineering:

- System Engineering
- Software Engineering
- Hardware Engineering

the Hardware Engineering is the fast-evolving part from Intel 4004 that had 2,300 transistors in 1970s to current AMD Epic Rome that has 32 billion transistors, which is about 10 million times improvement, in addition to that, the computer hardware's capacity and functionality all are increased significantly; the Software Engineering also has had huge improvement: there are so many programming languages, IDEs, tools. However, there has not been any significant change in the computing system engineering mechanism since the author got into the computer engineering 40 years ago.

At that time, the system engineering approach was the "waterfall", and current mechanism required in the ASPICE and ISO 26262 is the V mode that is essentially the same as the "waterfall".

Nowadays automation is used in almost every aspect of industrial control and development, but the system engineering is still at its original state done by manually, although some product development can be done using the tools like Simulink to do somethings automatically, the design mechanisms are still not changed.

The Data Driven System Engineering approach including the EPN notation in this book can be done recursively, which provide the possibility to automatically do the system engineering in the future under the help of the algorithms like AI that can figure out the relationships between the output and input signals in the system under development.

1.2.2 Clarify the System Development Process

It has been said: "the good beginning leads to the half success". The system engineering is the very beginning of every product development, and it lays out the infrastructure and the basic way or approach to develop the required products. If the suitable and efficient system engineering approach is chosen, then the development will be significantly improved. However, in most cases, the system engineering is the bottle neck in the development: every activity in it is done manually in sequence, different developers have different approaches to do the system engineering, the system engineers and the software or hardware engineers may have different understanding about the products; and in some projects, the development is done by literally following the processes or the standards without even understanding their meanings. In some cases, some activities are done even they are not needed.

The principle in every engineering is: Adding one more step in the engineering means adding one more chance to make mistakes in the product development.

The biggest obstacle in the system engineering is the engineers' mindset, which is the way that they are thinking. If the developers always follow the rules without questioning, then they will stay in the same way forever. If the developers don't know why to do the activities, then they won't be able to do them good, furthermore, they cannot do them more efficiently.

Fully understanding about reasons why the activities are needed is essential, especially when try to development something efficiently. So, we should always ask the two questions for each step, each component, each equipment, each person that is added in the development:

- What is the value for doing this step?
  - what do we expect from this step?
  - is it helpful for the development?
- If the answer is "yes" to the question above, then is there a way to optimize this step, such as combining it into other existed steps?

Furthermore, the way to break the obstacle in the system engineering is to find out the logical relationship between the system engineering activities by questioning, such as:

- What is the purpose of the Requirement Engineering?
- What is the purpose of the System Concept Design?
- What is the purpose of System Functional Allocation Design?
- What does the reliability mean?
- What does the Safety mean?
- What does the Cybersecurity protect?
- How do they above serve each other and in what way?

This book provides the author's understanding about them, and clarifies some vague points in the system engineering, especially in the automotive ECU engineering and in some industrial standards, Simplifies the engineering processes to make them more practicable and more efficient in clearly and completely defined processes.

For example, in ISO 26262, in the first step development: concept phase, the safety requirements are derived from the hazard analysis, which is confusing, because the safety requirements of every component in a vehicle must be derived from the vehicle safety requirements. And indeed, the approach of concept phase in the Part 3 of ISO 26262 derives the safety requirements indirectly from the vehicle safety requirements, as well.

Based on the safety criteria set in Table 3.4-15 Hardware Fault Metrics of ISO 26262, the author found that some of safety of autonomous driving systems in the current market are quite questionable. Currently the object detection systems used in the autonomous driving system either level 1 or level 2 have the confidence rate: 95% ~ 97%, which is: the best case that the system can detection the required objects is 97 correct cases out of 100 detections. Even if the object detection system could reach the 99% confidence rate, then that is 1 mistake out of 100 detections, and it is reasonable to assume that a vehicle will meet 100 objects in an hour, then comparing which with the PMHF required by the ASIL B in hardware criteria, that will be 10 million times difference, i.e., the current object detection system must improve 10 million times to meet the ASIL B criteria.

In the ISO 21434, the cybersecurity threats have the impact to the product operation, which can be covered by the either safety impact or the financial impact. So, there is not the necessary to address the operation impact.

in ASPICE, the interface work products are required between the development phases, such as that the system design specification is required to be the interface between the System Architecture Design and Software Requirement Engineering, which is not always valid, because there should not be any obvious boundaries or separation between the system development and the software development in a computing system development.

The engineering approach in the book covers every activity and the middle work products in the industrial controller development, especially by focusing on the automotive ECU development. For each activity in the development, the author would try to explain the purpose and what are the connections between the current one with others.

1.2 Intention

1.3 Intended Readers

System engineering is a logically complete process to development the products, in which, all activities are connected with each other from the requirement elicitation down to the system verification; and there is not significant difference between the system engineering and the software or hardware engineering, among which, the only difference between them is that the software or hardware engineering is the detailed level system engineering, and the system engineering is the higher level engineering from a specific development point of view.

So, to be a system engineer, it is better to know the whole "picture" of required product including the software engineering and hardware engineering, so that the purpose and meaning of each step can be captured to make better products.

However, in most organizations, the engineers are assigned to a very specific task in the development, such as:

- A Requirement Engineer
- A system Architect
- A Safety Engineer
- A Cybersecurity Engineer
- A System Integration Test Engineer or System Verification Engineer.

Nevertheless, to all readers:

- there is no right or wrong way to do the system engineering, there is only either the efficient or inefficient way to do it, which highly depends on the specific project.
- the final development target is to make the required product(s), the system engineering is only the facility to approach the target. So, the criteria of system engineering are if it can develop the product(s) efficiently.

To suit such reality, the following sections are the suggestions for the potential readers.

1.3.1 Industrial Electronic Control Engineers

An automotive electronic control unit (ECU) is a typical industrial electronic controller like the temperature controller in the air conditioning, the water flow and pressure controller in the city water system, motion controller in a robot, etc., so, all the activities in the automotive ECU development are applicable to every industrial electronic controller development.

However, three facts make an automotive ECU special:

- Massive Production: About 92 million vehicles were produced in 2019, such massive production makes the quality of all parts very important because if any defect causes the recall, then it will be huge financial impact to the business.
- High Safety Requirement: An automotive is a metal box that weights a few thousand kilograms and can have more than 100 KM / H speed, which make it very dangers, not only to the people inside of it, but also to the people on its path.
- A vehicle is the second expensive asset in most people's life, just after the house.

So, the automotive industry has the high standard requirements to an automotive ECU about the following three aspects:

- Quality
- Safety

- Security

which makes the automotive ECU development processes valuable to other industries.

The section of 2.2 Automotive ECU System Engineering Characteristics describes the brief and the complete overview about an automotive ECU development including:

- Basic Structure including the ECU network in a vehicle, a typic ECU and a typic microcontroller in an ECU.
- Standards about the safety and cybersecurity
- Quality processes
- Automotive OEMs in North America
- General development approach.

The detailed development methods for each development phase are described in the Chapter 3, which covers all the automotive ECU development activities from the requirement elicitation down to the system verification. Among which, all development activities are highly applicable to other industrial electronic controller development except the following aspects that are automotive specific:

- Feature Function
- Diagnostic Service
- Serial Communication in the vehicle network
- Cybersecurity

1.3.2 New System Engineers

The system engineering is a simple and straightforward process, at least, it should be, though the target product may be complicated.

If the readers understand how the "Hello World" works, then they got the essentials that a system engineer needs.

The chapter of "System Engineering Overview" gives the basic and essential picture about the system engineering, especially about the automotive ECU system engineering characteristics including the OEMs features, so it will help you to get into the real work environments.

The section of 2.2.1 Basic Structure has 5 layers structure about an automotive ECU in a vehicle:

- Vehicle ECU Network illustrated by Figure 2.2-1 Vehicle ECU Network, which describes a typical vehicle network topological structure and how the ECUs in a vehicle are connected each other.
- ECU structure illustrated by Figure 2.2-2 ECU Input and Output Signals, which described a typical automotive ECU structure including all the devices in an ECU.
- ECU software structure illustrated by Figure 2.2-3 Automotive ECU Functionality, which describes what typical functional components in an ECU focusing on the application layer components.
- Microcontroller in an automotive ECU illustrated by Figure 2.2-4 Microcontroller, which describes a typica microcontroller structure in an automotive ECU including the most common devices in it, such as RPU, APU, GPU, SHE, FPGA, RAM and ROM.
- Processor in a microcontroller illustrated by Figure 2.2-5 Security Hardware Extension, which described the structure of Security Hardware Extension in a

microcontroller that is similar with other processor or cores in a microcontroller.

The chapter of "Data Driven System Engineering" gives the full picture of what the industrial controller development, especially the automotive ECU development contents are, i.e., every automotive ECU development must have those steps, based on which, the readers can expend to the specific required development. Among which, there two components that are special:

- Feature Function, which is the most valuable part in an ECU.
- Application Mode Manager, which is the most important part in an ECU, which is the "administrator", for which, the section of 3.4.2.2 Application Mode Manager describes how an ECU operates internally and in a vehicle network.

After each section, there is the dedicated check list that are collected in Appendix A: Check List, which can help to quick cover the contents.

Don't be scared and don't be confused by so many regulations, standards, and steps. Try to avoid getting into the complicate and unnecessary steps and try to take approach directly to the target point.

Don't be limited or constrained by the book or other processes, the valuable character of a system engineer is Open Mind.

1.3.3 Veteran System Engineers

The book introduces a straightforward system development process together with a State Machine alike approach to develop the system driven by the data, both of which can be executed recursively.

The purpose is:

- Try to simplify the development process,
- Try to make the development process automatically.

If the logic relationship between the input signals and output signals can be figured out in a computer system using a database or transformation formula lists or AI for certain systems, such as automotive ECUs, then the development can be significantly improved, even furthermore, can be done automatically in the future.

Meanwhile, this book provides a full range of system engineering development activities, each of which has different approach from the conventional way of development, in some cases even different from some ISO standards.

For example, in ISO 26262, in the first step safety system development: concept phase, the safety requirements are derived from the hazard analysis. However, from the author point of view, the safety requirement development in every product should follow the generic system requirement development process, i.e., a product component requirement should be derived from the high-level product, so, an automotive ECU safety requirement should be developed from eventually the host vehicle. And indeed, the approach of concept phase in the Part 3 of ISO 26262 derives the safety requirements indirectly from the vehicle safety requirements, as well, and such "indirect" development makes that step confusing. So, in this book, an ECU safety requirement will be derived directly from the high-level safety requirement by the decompositions and derivations.

Another example is that the cybersecurity threat analysis gets rid of the "operational impact" that is required in ISO 21434, since it can be covered by either the financial impact or safety impact. And there are some other topics that the book has different understanding from some standards, which is based on the engineering principle: Adding one more step in the engineering will introduce one more chance to make mistakes in the product development.

Regarding to the reliability and safety development, this book provides the clearly defined scope and approach, which clearly defines that there are only two types errors in a computing system: Data Value Error and Data Timing Error, for which, the methods of complete coverage are described in 3.4.5 FMEA and 3.4.6 Safety.

1.3.4   Project Managers

There are three goals in the project management:
- Development Cost
- Product Quality
- Delivery on time

Those are exactly the goals for system engineering. The only differences between system engineers and the project managers are:
- System Engineers pay attention to technical aspects
- Project Managers pay attention to project budgets

This book tries to make the project managers understand how the system engineers' approach achieves the goals, so that the project managers can take the development activities under management.

For each electronic controller or automotive ECU development phase, there is the dedicated check list that are collected in Appendix A: Check List, project managers can take which as the schedule steps.

1.3.5   Software and Hardware Engineers

From engineering point of view, there is not any difference between the system engineering and software or hardware engineering. The only difference, if any, is that how or from which angle to look at it.

For example, the famous peace of source code: "Hello World": It should be a very simple software code from the programming point of view, however, it will be a very complicated software system from the compiling process or a huge system from the microchip point of view, which involves all the elements, such as CPU, RAM, Input device, Output Device, interconnect data bus, etc. to realize such peace of sentence.

So, the methodology in this book is also for the software and hardware engineering, and the software or hardware engineers will possibly figure out the better way to develop the parts if they look from the whole picture, because the relationship between the system engineering and the software or hardware engineering can be captured. On the other hand, the approach to do the system engineering is the exact same way to do the software or hardware engineering, as well.

The distinguish between the system engineers and the software or hardware engineers comes from some management, which assigns the specific tasks, such as software programming, hardware circuit design, to the dedicated persons, so that the dedicated persons are more familiar with certain activities in the development to make the management somehow easier. The shortfall of which is that those dedicated persons will lose the sight about the whole development and logic relationship between the activities, then if the development process is changed, or the position or work environment is changed, then they will have difficulties to capture up.

From business point of view, assigning the specific tasks to dedicated persons is not beneficial for the long-term run, because it is hard for the development team with the narrow sight to make any changes.

1.3.6   Verification Engineers

The system verification is part of system engineering, so the system verification engineers and the system engineers are same.

A huge and obvious misunderstanding about verifications in the automotive ECU development is that: there are always two teams at the system level development: one is the system engineering team which is responsible for requirement engineering and system architecture design, another is the system verification team to do system integration verification and system black box verification in most of OEMs and ECU suppliers, which is intended to follow the principle: the design and test should be done by the separated personals, so that they are independent each other to check the development from different angle, and to achieve the better results.

If the separation between the system designers and the system testers is to pursue the "independency" in the development, which is required by some standards, such as the Table 1: "Table 1 — Required confirmation measures, including the required level of independency" in Clause 6.4.9 of ISO 26262:2018-Part 2, then it is a misunderstanding: the independent confirmation required in the Table 1 above is about the "confirmation review", not about the test.

There are the critical and mandatory prerequisites to have such separation or independency:
- The designers and the verification testers must have same understanding about the products, though they may have different approach about the design.
- They must have same degree and same extent knowledge about the product development including the product operation environment.

In some special cases, those prerequisites are met, such as the verifications of ISO 26262 compliancy, AUTOSAR products, because those verifications are based on the well-known standards, and it is practical and feasible for all developers to have same understanding about them. And other cases where those prerequisites are met are in the specifications from the OEM to the supplier, because the OEMs' requirement specifications have been established for a while, so the verification team in the OEMs have the good knowledge about them. But in the cases of new products, especially in the suppliers' new product development, there are hardly the two teams that meet the prerequisites, and if the two teams don't meet the prerequisites, then the development result will:
- Reduce the development quality

- Increase the development cost and time

The reason is obvious:

If a system tester does not know the system design, then that tester will not be able to know what to test, which is, by the way, exactly the same to the designer: If a system designer does not know how to test the design, then that designer will not be able to do good design.

So, the better choice is to use the same team to do both the design and verification. If there must be two types of persons: Person type A and Person type B to do the system design and the system test, then the right way is:

- First, Person type A does the system design, Person B does the test case design.
- Second, Person type B does the system design review, Person A does the test case review.

The right time to design the system test cases is when the system is been designed.

The good way to design the system is to design both system and the test cases together, then to check the design from the test case point of view.

Development Principle: if a design cannot be tested, then the design is not trustable.

The verification or testing engineering goal is to use the minimum test cases to cover fully the system's functionalities.

In this book, the data in the system description: Output = f(Input, Middle Result 1, … Middle Result) represents fully and merely the system functionalities. So, as long as the data in the system description are based to build the test specification, then the goal is fully and efficiently achieved. And the data in the system description are all necessary for the system, so the test cases based on them are the minimum.

1.3.7 Safety Engineers

Being a safety engineer, especially being a safety engineer in the automotive industry, the coverage by the safety according to ISO 26262 and ISO 21448 is much more than what are required in the system engineering by the ASPICE.

The safety required in ISO 21448 needs the developments of safe driving situations of Safety of The Intended Functions (SOTIF), which needs to handle not only the systems' internal faults by following ISO 26262, but also the situations where the systems should avoid the misuses by operating outside the performance limitations.

The safety required in ISO 26262 and ISO 21448 are valuable topics not only in the development phases, but also in the production, maintenance, etc., this book only focuses on the safety in the development phases, which cover the three aspects:

- Reliability
  - Failure Mode and Effect Analysis (FMEA)
  - Fault Reaction (Safe State)
- Functional Availability
- Quality

Reliability means that the product acts in the way as implemented. Taking the famous "Hello World" software code as an example: If the code will always output the sentence: "Hello World", then it can be said that it is reliable because it does what is implemented. If there is a typo or mistake in the programming that wrote "World" as "Word", and if the code will always output the sentence: "Hello Word", then it can still be said that it is reliable because it does what is implemented, as well.

There is not such guarantee that a piece of software code will always do what it is implemented, because a modern microcontroller can run as fast as about 1,000 million instructions per second, which can be interfered by the electromagnetic noise, device defects, etc., the result of which may change the outcomes of the software.

To achieve the reliability, first it is to detect the potential errors, then design the reactions to them. In this book, the FMEA is totally based on the data in the system description: Output = f(Input, Middle Result 1, … Middle Result), since they represent fully and merely the system functionalities. So, there is not the need to develop the separated and dedicated function nets, and furthermore, the critical data can be easily and clearly figured out, so the protections and functional availability can be developed accordingly. The book provides the solution to fully detect the errors in the system under development, which consist of only two types of errors: value error and timing error.

The book also provides the evidence to meet the both ISO 26262 and ISO 21448 requirements, which can be a leading example for similar projects to achieve the compliance.

Meanwhile, this book introduces a script-based Freedom From Interference (FFI) detection method, which is based on if the Executable Procedure Node (EPN) defined with a Safety Integrity Level (SIL) modifies one or more data defined with higher Safety Integrity Level (SIL).

1.3.8   Cybersecurity Engineers

This book provides the approach to achieve the cybersecurity in the automotive Electronic Control Units (ECUs) that are inside of the vehicle network, which satisfies the UN ECE 155 regulation, the implementation contents are:

- Trusted contents in the ECU
- Authenticated access to the ECU
- Authenticated communication with the ECU

The book gives the explanations about why those implementations can satisfy the UN ECE 155 by listing all the items that are specified in the Annex 5: "List of threats and corresponding mitigations" in the UN ECE 155, which covers all the technical requirements in it.

The UN ECE 156 is about the vehicle software updates and update management system, all the detailed requirements in this regulation are documented in the chapter 7 "General specifications", which does not mainly concern the ECU development, rather, it is mainly about the ECU development organization processes and the updating procedures, so it is out of the scope of the book. However, to facilitate the UN ECE 156 compliance, this book provides the brief descriptions and suggestions about it in 3.4.7.5 UN ECE 155 / 156 compliance.

1.3 Intended Readers

1.4 Terminology

In this book, there are following special terminology in Table 1.4-1 Terminology below:

Terminology	Meaning
Arithmetic Logic Unit (ALU), Core, Central Processing Unit (CPU), Computer, Microcontroller, Processor	The Central Processing Unit (CPU)that sometimes is called as processor, processing core or core, consists of the Arithmetic Logic Unit (ALU) that does the operations, such as mathematics, logic operations, control registers, internal RAM and ROM, and peripherals. Both a computer and a microcontroller have one or more CPUs in them, a daily used computer has the monitor, keyboard, mouse, etc. that a microcontroller does not have, but can be connected to them by adding the HW design.
Error, Failure, Failure Mode, Fault	All those terminologies have the same meaning that the system does not act exactly as implemented.
Freedom From Interference (FFI)	An aspect of functional safety is designing the system to isolate safety-critical functions from other functions and ensure they are free from interference, which is required by ISO 26262.
Failure Mode an Effect Analysis (FMEA)	It is a method for identifying potential problems and their impact, which is required by ISO 26262 in the system design phase.
Mode Management, State Management	Both have same meaning, which is to control the running state.
Cybersecurity Security	Both have same meaning, which is to protect the contents from unintended access.
Test, Verification, Validation	They all have same meaning that the activities check the developed products' behaviors against the requirements and design specifications.

Table 1.4-1 Terminology

1.4 Terminology

2 System Engineering Overview

2.1 System Engineering Definition

- What is System Engineering?
- What is the purpose of System Engineering?

System Engineering is the approach to develop the required product (s), which includes three critical elements:

- The Target or Destination: the required product(s)
- The Approach: the way of making the product(s)
- The Starting Point: what is the current development situation, i.e., what are the current available resources

The three critical elements of system engineering can be illustrated as Figure 2.1-1 System Engineering Approach below:

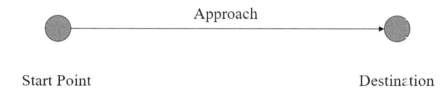

Figure 2.1-1 System Engineering Approach

The purpose of system engineering is to figure out the shortest path from the start point to the destination. It is obvious that it should be the straight line between the two points, however, it won't be so obvious in the real product development illustrated in Figure 2.1-2 Product Development below:

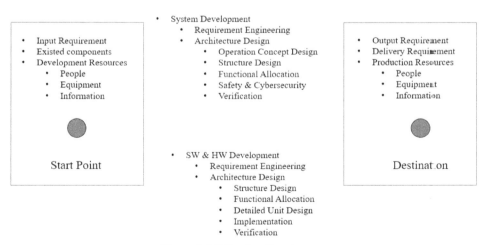

Figure 2.1-2 Product Development

What would be the shortest path from the Start Point to the Destination in Figure 2.1-2 Product Development above?

If the development path is following the V model specified by the ASPICE that is starting from the system requirement engineering down to the SW & HW unit implementation, then backing up from the SW & HW unit verification to the system verification, it is the normal way to develop products from scratch. However, none of development is from scratch, rather, every development is based on something that existed already, which will change the development path.

And considering the situations where there are the conditions for the start point and the destination, it is obvious that the approach should be different if the Start Point is different even for the same Destination. Figure 2.1-3 System Engineering Approach with Different Starting Point below shows an example that has three different approaches based on three different Start Points for the same Destination, in which the starting point represents the input requirement, existed components, development resources including the people who many have different level knowledge, equipment such as the software development integrated development environment (IDE), information such as whether the development information, background information, customer information distribute to the whole team, every of which will impact the development to result in different development path, which will change the development schedule.

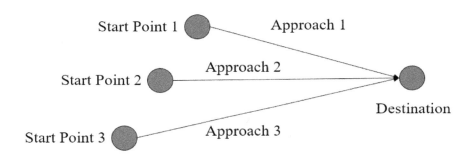

Figure 2.1-3 System Engineering Approach with Different Starting Point

However, in the reality, it is very common that the project's approach is always same no matter what the exist resources are.

So, it is worth to point out: even for the same required product(s), the system engineering contents (i.e., the system engineering approach) will be different if the existed situations are different even in the same organization. And even in the same organization, the approach will be different if the development team members have different experience or knowledge.

In another words, the system engineering is a "Dynamic" process.

So, for the same or similar required product(s), the better development way is to re-use the existed product(s) and knowledge to make the development less "dynamic", which involves two aspects to be reused:

- "Hardware" – such as the physical equipment, development environment, source code and data, which is the organization's property, and it is easy to be reused.
- "Software" – such as information, experience and knowledge (Human Resources), which is difficult to be reused, because it is kind of "dynamic" asset: people will leave, information will change, experience will be adapted to new development, etc., and it is critical part of reuse strategy.

The reuse strategy is to establish the engineering processes to keep all reused assets available, especially the information, knowledge and experience. In the automotive industry, the re-use strategy is implemented at all the levels:

- Organization Level
- Project Level
- Product Level

Those re-use strategies will be introduced at the last section of 2.2.7 System Engineering Approach of this chapter.

2.1 System Engineering Definition

2.2   Automotive ECU System Engineering Characteristics

The system engineering approach in this book is suitable to all the industrial electronic controller development, and it will mainly focus on the automotive electronic control unit (ECU) development which has the following characteristics in it.

2.2.1   Basic Structure

There are many "computers" in a vehicle nowadays that operate the doors, windows, brake, engine, radio, CarPlay, autonomous driving, etc., similar to the computers in an office and a family, those automotive "computers" that called automotive Electronic Control Unit (ECU) are connected each other illustrated in Figure 2.2-1 Vehicle ECU Network below to exchange information to collaborate.

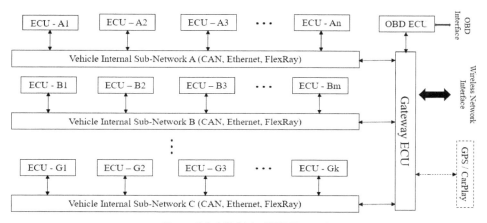

Figure 2.2-1 Vehicle ECU Network

An automotive ECU commonly has those functional parts that can be described as Figure 2.2-2 ECU Input and Output Signals below:

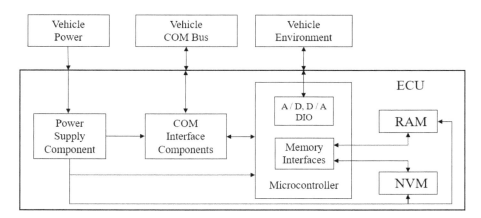

Figure 2.2-2 ECU Input and Output Signals

- Power: all electronic devices in an automotive must be powered by the battery in the vehicle, and the power path from the battery to the device or ECU is either permanently connected all the time or switched on and off by the vehicle ignition switch.
- Communication: it is the communication channel between the ECU and the host vehicle. Every ECU needs to cooperate with others in the vehicle to realize the functions. Some common communication protocols used in a vehicle are below, and all of which are serial communication protocols:
  - CAN
  - FlexRay
  - Ethernet
  - LVDS
- Environment Signal:
  - Temperature
  - Environment (Object detection) represented by either radar radio signals or by camera optical signals or ultrasound signals
  - Digital Input: such as ECU position signal
  - Digital Output: such as driving LED signal
  - Analog Input: such as fuel level signal
  - Analog Output: such as camera heating output voltage
- NVM: Non-Volatile Memory, that is a type of computer memory that can retain stored information even after power is removed. The NVM is divided into internal NVM that is allocated inside of microcontroller and the external NVM which typically refers to storage in semiconductor memory chips that are outside of the microcontroller, such as NAND flash and solid-state drives (SSD) chips.
- RAM: Random-access memory, that is a form of computer memory that can be read and changed in any order, typically used to store working data and source code. There is some amount of RAM inside of the microcontroller, however, to increase the system performance, almost all of ECUs need to use the External RAM to increase the memory capacity.

An automotive ECU commonly has following functionalities illustrated in Figure 2.2-3 Automotive ECU Functionality:

- Feature Function: The Feature Function is the only functionality that is needed by the host vehicle, which represents the required characteristics for the specific ECUs, such as: how the radar ECU detects the object(s) from the radio signals that are transmitted from the ECU antennas and reflected from the environment back to the ECU; how the camera ECU detects the object(s) from the environment that it can "see" through its lens; how the brake ECU controls the braking motor to output the braking force based on the brake pedal, vehicle velocity and trajectory. To realize the feature function, the listed below supportive functions are needed.
- Application Mode Manager: The Application Mode Manager is the component that manages all the components' operation mode in the application layer, and responsible for the whole ECU's safety management though the specific safety measures are implemented at the dedicated locations.
- Diagnostic Services: The Diagnostic Services are responsible for all maintenance activities, such as information query, parameter update, software content update,

service routine execution. The implementation is standardized though the parameters and configurations are project specific.

- Cybersecurity Function: The Cybersecurity Function provides the protection against the external threat and attack to the ECU's internal information regarding to the Authenticity, Integrity, Confidentiality, Privacy.
- Serial Signal Manager: The Serial Signal Manager is to process the serial communication signals between the subject ECU and the host vehicle to communicate with each other about the working status, and to receive and respond to the diagnostic requests that are needed by the Diagnostic Services.
- AUTOSAR: It is a software component package that provides the necessary services and devices management to support the ECU operation, which simplifies the ECU development.

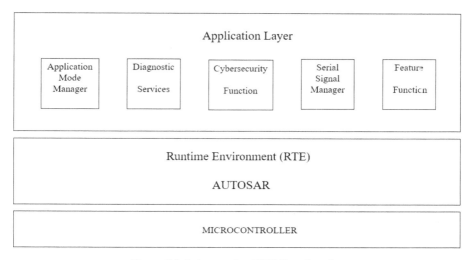

Figure 2.2-3 Automotive ECU Functionality

The key electronic part in an automotive ECU is the microcontroller, and a typical microcontroller electronic structure in an automotive ECU is like the one in Figure 2.2-4 Microcontroller, in which there mainly are:

- RPU: Real-Time Process Unit that is used to run the safety operations.
- APU: Application Process Unit that is used to run the application operations.
- GPU: Graphic Process Unit that is used to run the graphic operations.
- FPGA: Field Programmable Gate Array that t is an integrated circuit that can be programmed on the field to work as per a processor.
- SHE: Security Hardware Extension that is a processor dedicated only for cybersecurity operations.
- EEPROM: Electrically Erasable Programmable Read-Only Memory that is a type of non-volatile memory to store data or instructions by allowing individual bytes to be erased and reprogrammed. Nowadays large amount of EEPROM is replace by the Flash memory that acts as the EEPROM except the unit to be erased and reprogrammed is block, usually 128 bytes.
- RAM: Random Access Memory that is used for temporarily storing instructions

and data values.

- Interconnect: that is the communication bus connecting all those devices and processing units in a microcontroller.

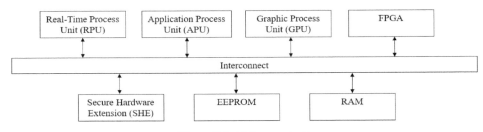

Figure 2.2-4 Microcontroller

Among a microcontroller, the Security Hardware Extension device is special, which needs to be paid attention during the microcontroller selection.

A typical Security Hardware Extension device is illustrated in Figure 2.2-5 Security Hardware Extension, which is just like a common microprocessor that has its own Arithmetic Logic Unit (ALU), ROM, RAM. What make it special are the cryptographic keys that are used to calculate the encryption and decryption by the cryptographic algorithms like the Advanced Encryption Standard (AES), Data Encryption Standard (DES), Rivest-Shamir-Adleman (RSA), Elliptic Curve Cryptography (ECC).

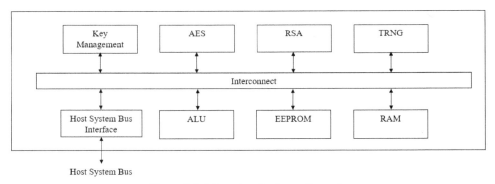

Figure 2.2-5 Security Hardware Extension

To ensure the security, once the keys are programmed into the ECU, no one is allowed to disclose them in any situation, neither the ECU developers nor the chip manufacturers. To use or check them, the users have to use the keys' IDs along with the operated data. The same security requirements are applicable to the True Random Number Generator (TRNG) and other relevant information, such as the keys' counters that prevent the keys from repeat use attack.

All the functionalities of Security Hardware Extension device can only be accessed using the dedicated instructions that are provided from the microcontroller supplier, and the device contents cannot be access by any other device in anyway, so, the ECU developers cannot develop any functionality if the device does not provide in the built-in.

From the industrial automatic control point of view, an automotive ECU is just like other industrial automatic controllers, such as temperature control in the air conditioning, water flow and pressure control in the city water system, motion control in a robot, etc., however, three facts make an automotive including all parts in it special:

- Massive Production: About 92 million vehicles were produced in 2019, such massive production makes the quality of all parts very important because if any defect causes the recall, then it will be huge financial impact to the business.
- High Safety Requirement: An automotive is a metal box that weights a few thousand kilograms and can have more than 100 KM / H speed, which make it very dangers, not only to the people inside of it, but also to the people on its path.
- A vehicle is the second expensive asset in most people's life, just after the house.

So, there is the significant impact to the people's safety and finance if a vehicle is out order, which requires the high standard requirements to the parts in a vehicle about the following three aspects:

- Quality
- Safety
- Security

2.2.2 Quality

A product quality means that the product should function exactly as the requirements:

- o The product design should strictly follow the product requirements
- o The product implementation should strictly follow the product design

An automotive is a very complicated product or system, in which there are about 30,000 parts, among them, there are about 80 ~ 150 Electronic Control Unit (ECU) components, one example of which, the Electronic Brake Control Module (EBCM) has about 30 functional modules, such as Vehicle Dynamic Control module, Electronic Parking Brake module, Service Brake module, etc., and millions software sentences inside.

To develop such complicated system, the ISO/TS 16949 is created to define the quality management system requirements for the design and development, production and, when relevant, installation and service of automotive-related products.

The proven better and efficient way to develop a complicated system like a vehicle is to use the modularization, i.e., to divide the complicated system into many simpler sub-systems, then each group of developers is taking care of one of them. To do so, the definitions of each sub-system contents and the interfaces between the sub-systems are critical.

For the mechanical parts, it is easy to define those interfaces and contents, such as the parts' shape, material, etc. because there is very clear boundary between the parts, there are fewer choices to the material and fewer ways to develop, and there is the very long history of their development.

However, for the computing software development, due to the special characteristics, that is: even for the same functional requirements, the different software developer will have different solution and different realization. So, it is infeasible to completely define the software interfaces and the contents, rather, that can be done only for some functionalities that are common for most ECUs, which is done by the AUTOSAR: AUTomotive Open System ARchitecture.

AUTOSAR not only simplifies the development, but also enhance the product quality by increasing the product reuse ability and by standardizing the components' requirements including the interfaces.

On the other hand, to regulate the development process, the automotive industry set up the software development process–ASPICE: Automotive Software Process Improvement and Capability dEtermination (also known as Automotive version of ISO/IEC 15504, or SPICE), which is a framework for software process assessment developed by the ISO (the International Organization for Standardization) and IEC (the International Electrotechnical Commission) in 1993. Its purpose is to evaluate development factors that allow assessors to determine an organization's capacity for effectively and reliably delivering software products.

- AUTOSAR is the modularization of product development
- ASPICE is the modularization of development processes

2.2.2.1 ASPICE

ASPICE requires that the organizations should have the established and managed development processes in place consisting:

- System Engineering Process illustrated in Figure 2.2-6 ASPICE
- Software Engineering Process illustrated in Figure 2.2-6 ASPICE
- Supporting Process including quality assurance, verification and configuration, etc.
- Management Process
- Reuse Process
- Process Improvement Process
- Supply Process
- Acquisition Process

The main target of ASPICE is to make the automotive software development reliable by dividing the whole development into multiple steps, and for each step, there should be the suggested indicators or work products to be checked.

- Advantage

The advantage to follow the ASPICE is to divide the complicated ECU software development into somehow simpler steps, so each step can be reviewed to make sure that it reaches the required quality target.

- Disadvantage

First, ASPICE requires that the whole development team has the same qualification, i.e., every team member should have the same understanding and same knowledge about the development, otherwise, the information will be misunderstood. Although it is true to every development, ASPICE extends such requirement further by adding more steps in the development, and adding more steps in a development means adding more development cost and adding more chances to make mistakes.

Second, due to the software development characteristics mentioned at the beginning of this section, it is very difficult to define the qualified interfaces or qualified indicators between those steps, so it may not be helpful to the development by adding more steps if the boundaries are not clear.

It is worth to point out: the interfaces for the mechanical parts in a vehicle have been established by so many part suppliers during so many years' development. So, there is few issues there. However, for a specific ECU software development, it is always done in a single supplier during only the limited time. So, it is infeasible to define the qualified interfaces and contents between those steps, which will compromise the development.

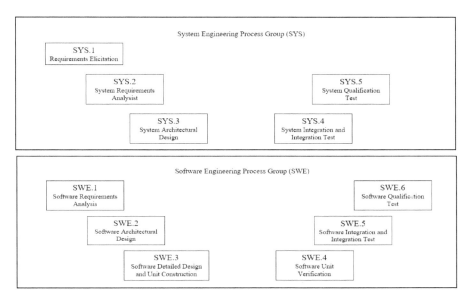

Figure 2.2-6 ASPICE

Last, ASPICE may confuse some developers who don't really understand the development by adding more steps. For example, in quite some automotive ECU development, the IBM DOORS is used to manage the requirements and design specification, and the IBM Rhapsody is used to design the architecture, which should be the design for the whole product from the system level down to the software implementation. However, to create the physical documents for System Design Specification that is the output work products from the step SYS.3 "System Architectural Design" in Figure 2.2-6 ASPICE above, the contents are exported from the Rhapsody by the system architect, then input into the DOORS as the Software Requirements, which, in turn, is imported into Rhapsody by the Software Requirements Analysis Engineers who work at the SWE.1 step: "Software Requirements Analysis". Among which, those exporting and importing:
- Does not add any value for the development
- Increase two chances to make mistake that may loss or misinterpret the information from the previous step.

2.2.2.2 AUTOSAR

As mentioned above, to simplify the automotive ECU software development, automotive industry developed the AUTOSAR: AUTomotive Open System Architecture illustrated in Figure 2.2-7 AUTOSAR below, in which, the contents between the Application Layer and the Microcontroller are the AUTOSAR modules, which are called

Basic Software Modules (BSW), and there is the full list of those modules released in AUTOSAR Release 4.3.1, which can be reused or adapted by most of automotive ECUs.

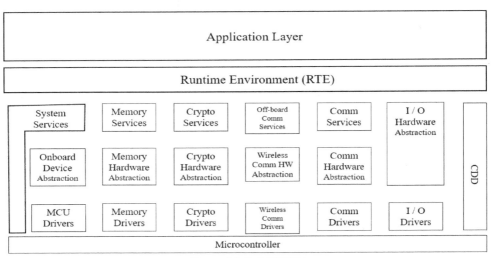

Figure 2.2-7 AUTOSAR

The fundamental concept in AUTOSAR is the modularization and separation of Application Layer from the AUTOSAR Basic Modules by defining the standardized contents and interfaces for each BSW and the interfaces between the Application Layer and the BSWs.

Based on such architecture, the ECU developers will mainly focus on the Application Layer development consisting of the application Software Components (SW-C) based on the selected microcontroller, all the rests belong to the AUTOSAR and can be purchased from the AUTOSAR BSW suppliers.

The BSW components in the AUTOSAR are divided into three layers:

- Driver Layer: MCU Drivers, Memory Drivers, Crypto Drivers, Wireless communication Drivers, Communication Driver Drivers, Input / Output Drivers
- Abstraction Layer: Onboard Device Abstraction, Memory Hardware Abstraction, Crypto Hardware Abstraction, Wireless Communication HW Abstraction, Communication Hardware Abstraction, Input / Output Hardware Abstraction
- Service Layer: System Services, Memory Services, Crypto Services, Off-board Communication Services, Communication Services

Note: In the BSW components, there are two special components:

- Input / Output Hardware Abstraction, which is directly interacted with the I/O drivers and the RTE to provide the functionalities, such as digital input signals like the ECU installation position, analog input signals like the temperature measurement, digital output signals like the LED driving signal, analog output signals like the camera heating voltage. Those functionalities are project specific, not common in the automotive ECUs, so there is not necessary to have the common service definitions. Those functions will be based on the microcontroller's peripherals, such as digital I/O pins, ADC / DAC, PWM, etc., which are common to the microcontrollers, so there is the driver definition.
- Complex Device Driver (CDD), which is for all the functionalities that have special purpose in the products, such as camera, radar, which are not common to

the automotive ECUs.

Based on the AUTOSAR structure, the product development is independent on the hardware, which is realized by following separations:

- The abstraction layer separates the service layer from the driver layer.
- The driver layer separates the abstraction layer from the microcontroller
- The whole AUTOSAR separates the Application layer from the whole hardware and service layers.

The benefits based on the AUTOSAR are:

- The ECU development is simplified significantly.

The development at the application layer does not need to handle the complicate hardware related aspects, such as the interactions with the drivers of MCU, memory, communication, etc., neither need to handle the communication between each application component, which will be done by the RTE that acts like the Windows in PCs.

- The application layer contents can be reused for different projects that use different microcontrollers.

Because the application layer does not need to handle the hardware related aspects, so it becomes hardware independent, which make the reuse possible.

- The development scalability is flexible

The ECU developers can select the scalability according the product specific functionalities and the safety requirements to configure the BSWs needed:

Category	Scalability Class 1	Scalability Class 2	Scalability Class 3	Scalability Class 4
OSEK OS	✓	✓	✓	✓
Counter Interface	✓	✓	✓	✓
Schedule Table	✓	✓	✓	✓
Stack Monitoring	✓	✓	✓	✓
Protection Hook		✓	✓	✓
Timing Protection		✓		✓
Global Time Synchronization Support		✓		✓
Memory Protection			✓	✓
OS Application			✓	✓

Table 2.2-1 AUTOSAR Scalability

2.2.3 Safety

In the automotive ECU development, the safety development is guided by both ISO 26262 and ISO 21448.

By following ISO 21448, the developers need to extend the safety mechanisms for both following vehicle operation scenarios:

- Ego vehicle operates outside the system performance limitations
- Ego vehicle operates within the system performance limitations

So that, the safety of automotive ECU systems especially the automated driving systems can be extended to cover more situations where not only the systems' internal faults but also the vehicles' misuses without any internal fault should be prevented.

The approach to do so in this book is:

- To identify the vehicle performance limitations and to avoid misuses by validating the systems' sensing and acting abilities against the road users and roads considering environment impacts such as geography, time of day, weather, infrastructure including the traffic signs.
- To develop the indicators (visual, auditory and haptic display) and the user instructions to warn the misuses once the ego vehicle operates outside the system performance limitations
- To development the internal safety mechanisms to prevent the systems from internal failures when the ego vehicle operates within the system performance limitations by following ISO 26262.

ISO 26262 covers not only the development including the concept development, the system development, the software and hardware development, but also the production, maintenance and services. In this book, the topics focus only the safety mechanism development including the concept, system, hardware and software development consisting of:

- Quality
- Availability
- Reliability

2.2.3.1 Reliability

Reliability means that the product acts in the way as implemented.

Note: The reliability is the compliance with the implementation, not with the requirements, in another words, if the product's output is compliant with the implementation, then the reliability is good. On the other hand, if there are the deviations between the implementation and the requirements, then it is the quality issue, not a reliability issue.

For example, the famous "Hello World" software code in C is like below:

```c
#include <stdio.h>
int main() {
  printf("Hello, World!");
  return 0;
}
```

If the code above will always output the sentence: "Hello, World!", then it can be said that it is reliable because it does what is implemented.

If there is a typo or mistake in the programming that wrote "World" as "Word" like below:

```c
#include <stdio.h>
int main() {
  printf("Hello, Word!");
  return 0;
}
```

And if the code above will always output the sentence: "Hello, Word!", then it can still be said that the code is reliable because it does what is implemented, as well.

If some readers think that the output result MUST be the one as implemented, then think again: A modern microcontroller can run as fast as about 1,000 million instructions per second, and an instruction in a 32 bit microcontroller consists of 32 binary bits of "1" or "0" like: 111000... 10, each "1" or "0" is called the binary signal and represented by an electronic signal state: high or low electronic voltage, and there are thousands mega such signals running at such fast speed in a microcontroller, if any of them goes wrong, then the result may not be the one as implemented.

So, the reliability is the ability to do what it is meant to do. Considering the situations about the binary signal complexity and speed described above, and the components in a microcontroller described in Figure 2.2-4 Microcontroller, and the environment described in Figure 2.2-2 ECU Input and Output Signals, there is not guarantee that a software product will always behave as implemented.

To achieve the reliability, the system needs to detect if there is anything wrong during the operation, and if there is, then the system either corrects it or sends out a warning, which is a simple and complicated topic.

It is a simple topic because there are only two types of errors that can occur in a computer including automotive ECU microcontroller.

All information in a computer or an automotive ECU microcontroller, either operation instruction or data, is represented by an electronics signal mentioned above, which can have only two types of errors:

- Signal Value error
- Signal Timing error

The signal value error is defined as that the signal value is not one that is supposed to be, which is, in turn, caused by two types of errors: transmission error and operation error. The transmission error happens during the information transmission from one location to another. For example: a signal: 1101 is transferred from location A to B, and the signal value is changed to 1100 during the transmission that may be cause by some electromagnetic glitches.

The operation error is defined as that the operation instruction is changed unintendedly. For example: a signal X is operated by an instruction: 1101, however, the instruction is changed to 1100 during the operation by some electromagnetic glitches, then the result will not be the one expected.

In a computer, the transmission error can be detected by the computer hardware using some technology such as: Cyclic Redundancy Check (CRC), Checksum, Error Correction Code (ECC), etc. The signal value error is the difficult issue that needs to be handled, which leads the following complicate topic.

It is a complicate topic because the binary information is everywhere in a computer: the memory addressing register, the intercommunication bus control register, the internal devices status registers, especially the information occurs in the Arithmetic Logic Unit (ALU) that is the computer's core, in which the information is expected to be changed. So, there is no way to tell if the change is intended or unintended.

So far, the only way to prevent such error is to use two computing processors to do the exact same operations for each step, then compare the results from every step, the operation result won't be trustable if there is any difference between the two processors. So, in the Real-Time Process Unit (RPU), there are always two exact same processors or cores that run in the "Lock Step" mode to do the same operation in the exact same way, then compare the results during in the operation, which is the reason why the safety tasks are always done by the cores that run in the "Lock Step" mode in the automotive ECU.

The signal timing error is that the signal does not occur at the required time. The detection to such error is divided into two parts: the contents inside of a processor of computer and the contents from outside of a processor of computer.

To the contents inside of a processor of computer, this type errors cannot be detected directly, neither by the computer nor by the application, because all the execution instructions in a processing core are executed in serial sequence by the arithmetic logic unit (ALU) based on the system clock, so the detection instruction and the under detected instruction or data run on different time, i.e., they don't have the common referenced time base to check the timing.

To the contents input from outside of a processor of computer, the time errors can be detected based on the input communication protocol.

2.2.3.2   Availability

The availability is the products' ability to provide the required functionalities even when something goes wrong. To do that, it requires that the product should have the redundant mechanism for certain important functionalities. For example, to detect the objects on the road, the autonomous driving vehicle will have at least two mechanisms for such object detection, one uses the radar, another uses the camera, and they are independent each other, so that the object detection ability is increased in such way in case where either the radar or the camera is out of order.

Another example is the braking system in a vehicle, which consists of two sub-systems: the Electronic Control Braking System (ECBS) that is the main brake system and the Electronic Parking Braking System (EPBS) that is the backup brake system, i.e., in the case where the main braking system: Electronic Control Braking System (ECBS) goes wrong, then the Electronic Parking Braking System (EPBS) can be used to brake the vehicle.

Having the redundant sub-systems in a product will increase the functional availability, in turn, which will increase the safety, as well. However, it will:
- Increase the cost to the product.
- Increase the complexity of the product, which will impact the product's safety and quality, and in some cases, it will decrease the product's availability somehow. Taking the object detection as an example which has two sub-systems: camera and radar. If the camera detected an object but the radar didn't, then the common autonomous driving adjudgment is to take it as there is an object, which will lead to the wrong result that the vehicle will stop driving if the camera was wrong, i.e., adding one more component in a product will add one more chance to make a mistake. The better solution to apply the redundant components is to use three same or similar components, then take the result from voting among them if there is any difference between them, well, that will increase the cost significantly.

From some points, system engineering is an art of balance.

2.2.3.3   Quality

There are too many instances in automotive industry that the automotive companies were out of business because their products have certain quality issues that caused the huge financial impact.

Quality control is vital, and computer software development is prone to mistakes.

Quality is the important part of safety, and the product safety is ensured by both technology and quality.

The technology provides the technical solutions to achieve the technical goals, such as reliability, availability that are described above. As mentioned in the section of 2.2.2 Quality, the quality ensure that the development strictly follows the product requirements:

- How to develop the accurate and qualified requirements for the required product is the topic of technology.
- How to avoid making mistakes in the development is the topic of quality.

Quality can be controlled and ensured by two aspects:

- Management: it is to ensure the quality from organization point of view. The organization needs to set up the development processes, especially the processes about quality assurance, such as ASPICE, ISO/TS 16949.
- Technology: it is to ensure the quality from development technology point of view, such as executing the system integration test, system black box verification.

2.2.4   Security

The security goal is to prevent the ECU from the unauthenticated interactions including modification, operation, misinformation and discovering.

The automotive ECU cybersecurity threats or attacks may apply the impacts by:

- executing a piece of code in the ECU to manipulate the vehicle, to do so, there are only two possible situations: either in the ECU production when the software is downloaded to the ECU, or during the updating software process where the new version software will be downloaded to the ECU.
- Accessing the ECU using the diagnostic services through the On-Board Diagnostic (OBD) connections or using the testing interfaces, such as JTAG, XCP, to modify the software contents or parameter values, or execute some service routines.
- modifying or interfering the communication with the ECU to manipulate the ECU's behavior.
- reversing engineering methods to retrieve the information from the ECU.

To most ECUs in Figure 2.2-1 Vehicle ECU Network, the first three threats in the list above are the development focus, the last one has very low attack feasibility according to the analysis based on the ISO 21434, and the application can do little about it, so it will not be discussed in this book.

So, to the automotive ECUs mentioned above except the Gateway and On-Board Diagnostic ECUs, the cybersecurity development contents are to prevent ECU from the first three attacks, for which, the measurements are to ensure:

- Trusted contents in the ECU
- Authenticated access to the ECU
- Authenticated communication with the ECU

To achieve the goal, there three aspects to be insured:
- The contents inside of ECU are authenticated or confidential.
- The access to the ECU is authenticated.
- The communication with the ECU is authenticated or confidential.

In the automotive industry, the security is governed by the UN ECE 155 / 156 which covers the EU countries, USA, Canada, Japan and South Korea.

Similar to the safety ISO 26262, the security standards require the developers to demonstrate two aspects to meet the security requirements:
- Good technology: the developers need to have the product design and implementation good enough to meet the required security requirements.
- Good faith: the developers need to demonstrate that the development is managed by the good processes that are required by the UN ECE 155 / 156.

This book will cover the all the topics related to the technology, not the management.

There are the following differences between the security and the safety:
- The safety is about the inside faults, the security handles the outside invasions.
- The safety is against the errors, the security is against the intelligent hackers.
- The safety makes sure that the product is safe only when it is used, the security secures the information inside the ECU both when it is used and when is not used, i.e., the information still needs to be secured even after its lifecycle.
- The safety does not consider the financial loss, the security must prevent the financial property from loss.

2.2.5   Various ECUs in a vehicle

In a modern vehicle, there are multiple ECUs in the vehicle that can be grouped into the following:
- Conventional ECUs
  - Chassis: Brake Control Module, Suspension Control Module, Steering Control Module, Parking Brake Control Module, etc.
  - Body: Body Control Module, Door Control Module, Gate Control Module, Seat Control Module, Window Control Module, etc.
  - Vehicle Network: Gateway Control Module, Wireless Control Module, etc.
  - Power Train: Engine Control Module, Transmission Control Module, Fuel Pump Control Module, etc.

- Advanced ECUs
  - Autonomous Driving: Radar Sensor Module, Camera Sensor Module, Automatic Emergency Brake Module, Blind Spot Detection Module, Line Keeping Control Module, etc.
  - Infotainment: Radio Module, Cluster Module, Human Machine Interface Module, etc.
  - Vehicle to X networking: Vehicle to Vehicle Connection Module, Vehicle to Infrastructure Connection Module, etc.

For each of them to work, it needs to communicate with each other to know the vehicle and other ECUs' status, which make the communicate very important.

2.2.6   Customer

In north America, the traditional big three automotive Original Equipment Manufactories (OEM) are below:

- General Motor (GM)
- Ford Motor (Ford)
- Chrysler Automobiles (Chrysler)

Regarding to the ECU development, they all have the common approach:

- Employ the ASPICE as the development processes
- Apply the AUTOSAR as the product development approach
- Take ISO 26262 as the safety guidelines
- Engage the UN ECE 155 / 156 as the Cybersecurity guidelines

On the other hand, from the suppliers' point of view, each one has its own characteristics:

- GM: ECU development is driven by the specifications
- FCA: ECU development is driven by the target
- Ford: ECU development is driven by the engineering methodology

2.2.6.1   GM

The only part that GM needs from the ECU suppliers is the Feature Function described in the section of 3.4.2.1 Feature Function, all other parts in an ECU must follow the GM specifications, and GM has not only the very detailed and specific functional requirements but also the design specification about the ECU development, which consists of mainly following types specifications:

- Implementation specifications, which are documented in the serious of GBxxxx, which includes:
  - AUTOSAR infrastructure
  - AUTOSAR configuration
  - Bootloader
  - Diagnostic services
  - Application Mode Management
  - Vehicle network communication
  - ECU software update procedure
- Cybersecurity specification (both software and hardware) CYSxxxx.
- Verification specifications, which are documented in the serious of CGxxxx.
- The ECU specific requirements are documented in the Component Technical Requirement Specification (CTRS).

The GM development advantage:

- The development will focus on only the required features that will be implemented at the application layer (SW-Cs), and the AUTOSAR must be implemented by GM's certified suppliers.
- The GM's requirements are very specific, so the ECU suppliers don't need to design much besides the required features.

The GM development disadvantages:

- The ECU suppliers must follow GM's instruction for every detail in the ECU

infrastructure development, so they don't have much flexibility.
- Most of ECU suppliers have already developed the products in their own way, then they must adapt the development to meet the GM's specifications, which will result in significant changes and costs, which does not have much value to either the ECU function or the vehicle performance.

2.2.6.2 Chrysler

Chrysler pays most its attention to the final goal of the required product. As to the development procedure and the detailed implementations, the ECU suppliers have much more flexibility comparing with GM.

For each ECU project, Chrysler provides two types of important specifications:
- PF_xxxx: Product Function specification, which specifies all the requirements about the ECU's infrastructure including:
  o Hardware interfaces
  o Communication interfaces
  o AUTOSAR specification including the diagnostics.
  o Engineering requirements
- VF_xxxx: Vehicle Function Specification, which specifies how the ECU should work in the vehicle system by focusing on the required features.

The Chrysler development advantage:
- The ECU suppliers have much more flexibility by mostly re-using what they have already developed.
- The development is more efficient because the supplier can do the development in its own way.

The Chrysler development disadvantage:
- Since Chrysler fully trusts the ECU suppliers with a little monitoring during the development, so the deviations between what are required and what are implemented are always found out at the final phase, which cause that the development is postponed and that the fix is already at the high cost.

2.2.6.3 Ford

Ford ECU development requirements are documented totally in the Engineering Statement of Work (ESOW) document which will be delivered to the ECU supplier at the project beginning, in which, there are three parts outstanding comparing with other two:
- Skills Matrix
- Functional Specification described using Hatley-Pirbhai method
- ECU Software Testing Requirements

Skills Matrix: Ford gives the clear requirements about the development team members from both Ford and the ECU suppliers, for example, for the Core Engineering Approaches, there are the following nine requirements:
- Experimental Methods
- Customer Focus
- Design Verification Process
- Process Control Methods

- FMEA Process
- Global 8D Process
- Reliability Methods
- Robustness Methods
- System Engineering Process

For each of above, there are the dedicated requirements for the three phases in the developments:

- Concept Phase
- Implementation Phase
- Manufacture Phase

And for each phase, there are three skill levels required:

- Awareness: Needs awareness on this skills area to understand what the subject is and to identify if further knowledge is required.
- Practitioner: Has attended necessary training and can apply to simple situations - needs help with more complex applications.
- Advanced Practitioner: Practitioner with sufficient experience to tackle all situations without assistance. Could facilitate others. He/she is not a Subject Matter Expert or a Specialist.

Based on the Skill Matrix, the project management will have enough information to prepare the development team members and provide the required training.

Functional Specification: Ford provides the required feature function requirements using Hatley-Pirbhai method, which is kind of State Machine method. In such way, the requirements will be described in clearer and more logic way.

ECU Software Testing Requirements: that is a short document, however, it is very critical one that provides the Ford detailed software test requirements based on the Functional Importance Classifications: Class C (most important), Class B (middle important), Class A (less important).
The document specifies what test coverage a specific class functions should have.

The Ford development advantage:

- The development requirements are clear, specific and logic.
- Advanced methods are used.

The Ford development disadvantage:

- It is difficult to fully meet the Ford development requirements. For example, according to one of author's Ford project experience, the power liftgate control module full software test could take about 10 years by a development team consisting of 10 software developers if the test would follow the Ford ECU Software Testing Requirements strictly.
- It sets high standards for both the customer and the ECU suppliers, which requires the whole development team members to have such knowledge and experience, which is hardly met in most development projects.

2.2.7 System Engineering Approach

In automotive industry, to achieve the goals of products' quality, safety and cybersecurity, the industry established some procedures to develop the required products, the followings are some of commonly used approaches on the organization level, project level and product level.

2.2.7.1 ASPICE

The ASPICE is the organization level approach to do the development.

All automotive OEMs require all the ECU suppliers to have the ASPICE compliant processes in place to develop the automotive ECUs, which is to ensure that there are the established procedures in the suppliers' organizations, so the development is regulated.

- Engineering Process, which is introduced in the section of 2.2.2.1 ASPICE
- Supporting Process, mainly including verification, quality assurance, configuration management
- Management Process, including project management, risk management and measurement
- Reuse Process
- Process Improvement Process
- Acquisition Process
- Supply Process

Similar to all other standards, the ASPICE provides the guidelines for the automotive organizations to develop the software products used in the vehicles, to achieve the best results from which is to adapt it into each specific situation, not just following it literally. For Example, ASPICE requires that the development uses the approach of Top – Down, however, almost none of automotive ECUs is done in such way, because the ECU suppliers must have a working ECU first in order to quote the customer projects, otherwise, they won't even have the chance for the development. Once the supplier is awarded the project from the customer, the supplier is expected to adapt the existed ECU to the customer project, rather than re-design the ECU from top to bottom.

2.2.7.2 Base Project vs Application Project

At the project level, the commonly used approach is the Base Project vs Application Project.

As mentioned at the beginning of this section, the automotive ECUs are massively produced products, so the most common and efficient way do the development is to develop the Base ECU first that have common required features, then adapt it to the different OEMs, different platforms and different model years.

This approach is the Base - Application project approach, which is used by all the OEMs and all the automotive ECU suppliers.

For this approach to work, the base project design needs consider:

- The common required features from all the customers
- The common interfaces from all the platforms.
- Adapting functions from the base project to the application projects by tuning the parameters rather than changing the source code.

The good example of such Base – Application project approach is the AUTOSAR, for which, the AUTOSAR components suppliers develop the base project, and all other users develop the application projects by adapting the AUTOSAR components to the required ECU platforms for the customers by tuning the components' parameters, or by configuring only needed components for the required scalability.

Another good example is the Electronic Parking Brake (EPB) system, which has well defined the functionality and interfaces by the VDA, and all relative information is well documented.

2.2.7.3 Development Based on AUTOSAR

At the product level, the approach is to develop the ECUs based on the AUTOSAR that is introduced in the section of 2.2.2.2 AUTOSAR, and nowadays almost all automotive ECUs are required to be developed based on the AUTOSAR except some long history ECUs, such as Electronic Brake Control (EBC), Engine Control Module (ECM), etc., the approach based on the AUTOSAR enables the developers to focus on only the application layer, which pretty much like the development under the Windows on the PCs, i.e., the AUTOSAR is the operating system in an automotive ECU.

Based on the approach, the development at the application layer does not need to handle the complicate hardware related aspects, such as the interactions with the drivers of MCU, memory, communication, etc., neither need to manage the communication between application components and the BSW components, so that the most of ECU components can be reused for different projects that use different microcontrollers except the feature functions that will be introduced in the section of 3.4.2.1 Feature Function, because the application layer does not need to handle the hardware related aspects, so it becomes hardware independent.

This approach can be easily extended to other industrial automatic control development, such as the robot control, industrial machine control.

2.2 Automotive ECU System Engineering Characteristics

3 Data Driven System Engineering

The goal of system engineering is to figure out the efficient way to develop the required product whose output signals meet the requirements consisting of the input signals, the environment conditions, the safety and cybersecurity requirements.

System engineering is a product development approach based on the experience, and there is not the absolute right or wrong way to do it.

Even for the same required product, different team has different ways to develop the product, even for the same team, there are different ways to do it under different circumstances.

In this book, the Data Driven System Engineering is the system development approach based on the data flow from the system input signals to the system output signals, which consists of the following steps:

- Understand what is required including:
  - o Requirements Elicitation
  - o System Requirements Analysis
- Design and implement the required system (System Architecture Design) including:
  - o System Operation Concept Design
  - o System Structure Design
  - o Electronic Architect Design
  - o Functionality Allocation
  - o System Failure Mode and Effect Analysis (System FMEA)
  - o System Safety Development
  - o System Cybersecurity Development
- Verify the system including the System Integration Test and System Test

3.1 Concept

3.1.1 Data Driven Development

A computing system can be illustrated as below from the external point of view:

Input Signal ⟹ Process (System Under Development) ⟹ Output Signal

Figure 3.1-1 Computing System

in which, the output signal is the outcome of the system under development based on the input signal, i.e., the output signal represents expected behavior under the input signal for the system under development, so the output signal represents the functionalities of the system, and the system development goal is to derive the output signal based on the input signal.

For example, in Figure 3.1-1 Computing System above, the output signal may represent the braking force that is the output from the Electronic Brake Control Module (EBCM), the engine rotation speed that is the output from the Engine Control Module (ECM), the detected objects from the Object Detection Module, etc., So, it can be said that the output signal fully represent the expected external behaviors of the system under development, and the goal of system development is to figure out how to derive the output signal from the input signal under the required environment.

The system in Figure 3.1-1 Computing System above can be described as a function between the input signal and the output signal as below:

Output signal = f(Input signal)

Then, the goal of system development is to:

- Define the output signal and input signal
- Figure out and implement the relationship between the input signal and the output signal that is represented by the function: f

In all real projects, the relationship between the output signal and the input signal is not such simple that it can be directly described as a function, rather, the output signal must be derived from the input signal via some middle results, so the system under development should be described as:

Output = f(Input, Middle Result 1, ... Middle Result n)

Furthermore, in the real product development, there are multiple output signals and input signals for a system or product, so the system can be illustrated as Figure 3.1-2 Computing System with Multiple Signals below:

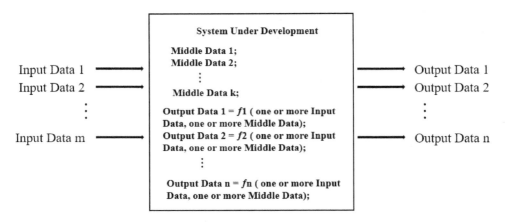

Figure 3.1-2 Computing System with Multiple Signals

So, the relationships between the data can be described as:

Output Data 1 = $f1$ (Input Data 11, ..., Input Data 1i, Middle Data 11, ..., Middle Data 1j);
Output Data 2 = $f2$ (Input Data 21, ..., Input Data 2l, Middle Data 21, ..., Middle Data 2p);

...

Output Data n= fn (Input Data n1, ..., Input Data nq, Middle Data 1n, ..., Middle Data nr).

Among the formulas above, m, n, k, i, j, l, p, q, r all are integers with the relationships:
$1 <= i, l, q <= m$; $1 <= j, p, r <= k$;
all the input data groups consisting of:

- the group of Input Data 11, ..., Input Data 1i,

- the group of Input Data 21, ..., Input Data 2l,
- ...,
- the group of Input Data n1, ..., Input Data nq

are subsets of the input data group consisting of Input Data 1, ... Input Data m; and all the middle data groups consisting of:

- the group of Middle Data 11, ..., Middle Data 1j,
- the group of Middle Data 21, ..., Middle Data 2p,
- ...,
- the group of Middle Data n1, ..., Middle Data nr

are subsets of the middle data group consisting of Middle Data 1, ... Middle Data k.

So, a computing system development can be described as:

- defining a plurality of data consisting of 3 types of data:
  one or more input data;
  one or more middle data;
  one or more output data;
- each data above has two, and only two attributes: data value and data timing.
- defining data calculations for each defined output data using one or more defined middle data and one or more defined input data, consisting of followings or combinations of following:
  - o mathematical expression
  - o logical expression
  - o fuzz expression
  - o experience expression
  - o AI expression
- designing needed hardware devices and software functions to realize those data and calculations.

Based on the descriptions above, the system or product development will be done by the following activities:

First, the development needs to define the input data and output data, and those data fully describe the required system from the external point of view or the system black box point of view. Which is context of the **System Requirement Elicitation and the System Requirement Engineering**.

Based on the system external descriptions using the input data and output data, the development needs to design and realize all needed middle data and establish the relationships: $f1$, $f2$, ..., fn to derive the output data from the input data via the middle data, the development process of establishing those relationships is the **System Operation Concept design**.

The system operation concept is the most essential concept in every system development, and none of system development can be done without such concept, because the required output signals cannot be figure out without the relationships.

Once the operation concept is established, the essential data flow logic from the input signal via the middle result to the output signals are known, so the following developments can be done:

- Establish the functions to derive those needed middle results from the input signals and other middles results.
- Allocate those middles results to the suitable devices in the system.
- Establish the data flow, data capacity and control time sequence for the middle data, input data and output data.

Those contents above are the **System Structure Design.**

Based on the data flow, data capacity and control time sequence needed by the data functions and storages defined in the system structure, the development needs to design the electronic element architecture to satisfy those requirements, which is the **Electronic Architecture Design.**

Once the needed electronic devices are in place, the designed functionalities can be allocated to the specific functions, partitions and electronic device, which is the **Functionality Allocation.**

Once the needed electronic devices are in place, and on which the dedicated data and data functions are allocated, the following information are known:

- How the data is transmitted and transformed
- What devices have what data

Based on the information above, the error analysis can be done about what potential errors could happen and where, which are the contents of **System Failure Mode and Effect Analysis (System FMEA).**

If the system under development is a safety relevant product, then the required safety level must be achieved, which will include:

- To achieve the required reliability, the system must development the mechanism to tolerate two types errors in the system: Signal Value Error and Signal Timing Error.
- To achieve the functional availability, the system should ensure that the required critical safety functions will be available at the required ratio time.
- To achieve the required quality, the development needs to follow certain requirements.

Those contents above are the development of **System Safety**.

For the modern vehicles, all automotive ECUs need to have certain cybersecurity features, especially for the vehicles that will be sold to the aeras where are governed by the UN ECE 155 / 156 regulations.

The ECU cybersecurity ensures authenticity and confidentiality of the ECU contents and the communication between the subject ECU and the host vehicle. Those contents are the development of **System Cybersecurity**.

Once all the functional elements are ready, those functional elements should be integrated together in the specified devices, and the outcomes from the integrated elements should be measured to against the design specifications, which is the **system integration and system integration verification.**

Finally, the developed system should be verified in the real required environment or simulated environment to against the required environmental and lifetime conditions, which is the **system black box verification.**

All the activities above are common to every industrial electronic control system development.

The features introduced in this book are to make use the data attributes in a computing system that are disclosed below and optimize the development activities.

All information in a computer or an automotive ECU microcontroller is represented by an electronics binary signal like 1110010, which represents only two types of information:

- operation instruction
- data

and both of operation instructions and data have only two attributes:

- Value, which is the quantity represented by the binary.
- Timing, which is the time when the binary is expected to occur.

among which, the operation instruction serves the transporting data from one location to another, transforming data from one form to another, and the data represent the signals' values:

- Output data represent the output signal values
- Middle data represent the middle result values
- Input data represent the input signal values

Base on the analysis above, it can optimize the development activities by focusing on the data only, specifically focusing on the data attributes: Value and Timing.

Let's see what the system operation concept that was discussed above is really about. As the discussion above, the system operation concept can be described as:

Output Data 1 = $f1$ (Input Data 11, …, Input Data 1i, Middle Data 11, …, Middle Data 1j);
Output Data 2 = $f2$ (Input Data 21, …, Input Data 2l, Middle Data 21, …, Middle Data 2p);

…

Output Data n= fn (Input Data n1, …, Input Data nq, Middle Data 1n, …, Middle Data nr).

Taking one of them as an example, which is the Output Data 1 derivation below:
Output Data 1 = $f1$ (Input Data 11, …, Input Data 1i, Middle Data 11, …, Middle Data 1j);

The development goal is to figure out the relationship or the function: $f1$ to derive the Output Data 1 from the Input Data 11, …, Input Data 1i via the Middle Data 11, …, Middle Data 1j. The relationship $f1$ not only derives the output data 1 value from used the middle data and input data, but also reveals the timing for the output data 1 to occur. which is based on the duration of $f1$.

The $f1$ may be a mathematic or logic expression, a fuzz expression, an experience or AI relationship, the development of which belongs to the specific project or product, not in the computing system engineering scope. The system engineering is to provide the reliable and working environment to execute the $f1$ and accommodate the needed data, which will be done by ensuring the information trustable in the system including both operation instruction and data.

3.1 Concept

Ensuring the operation instructions trustable has to rely on the hardware's functionalities, about which the application software can do little. So, it will not be the system engineering's focus.

Ensuring the data trustable is the system engineering main topic, and by focusing on which, it will optimize the system engineering activities in the following aspects:

- The system operation concept provides the correct paths and the perfect granularity of development activities, which results in persistent and full system coverage without any redundancy, because all the information in the concept is necessary to the development. If any data in the concept can be directly realized by the implementation, then it will be directly defined and used in the development; If any data in the concept needs to be figured out by further decomposition using a mathematic or logic expression, a fuzz expression, an experienced expression or AI derivation, then it can be treated as sub-system, for which all the development activities above will apply, which results in the recursive development process, which in turn provides the possibility of automatic development.

- The correctness of development, such as reliability, FMEA, safety, will apply only on the data that represent fully the system functionalities, i.e., if all the data in the concept are reliable, correct and safe, then the system must be reliable, correct and safe, which results in another level high efficiency.

- The system operation concept provides the persistent and defined development path for all the activities, which will share the development path to increase the development efficiency comparing with the conventional development, for example, in which the FMEA needs to development its own function nets, failure nets.

- The system operation concept provides the accurate integration interface definitions and the relationships. For example, the $f1$ provides the precise integration path, i.e., if any component's functionalities or interfaces are not according the concept expressed by the $f1$, then it will not be able to run in the function; if any component does not follow the concept and tries to be integrated into the system, then it will not be able to find its location.

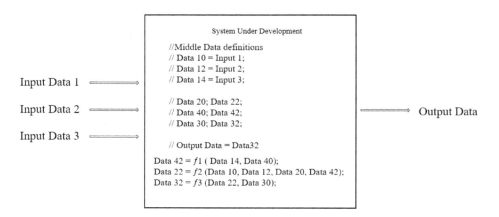

Figure 3.1-3 Example System

Taking the following system in Figure 3.1-3 Example System as an example to describe the development.

In the system above, the system operation concept is defined by the following relationships or formulas between the output data and the input data via some defined middle data:

Data 42 = $f1$ (Data 14, Data 40);

Data 22 = $f2$ (Data 10, Data 12, Data 20, Data 42);

Data 32 = $f3$ (Data 22, Data 30);

Among them, the Data 10, Data 12, Data 14 are the input data buffers, the Data 32 is the output data buffer, the Data 20, Data 22, Data 40, Data 42 and Data 30 are the designed middle data. The functions of $f1$, $f2$, $f3$ may be a mathematic or logic expression, a fuzz expression, an experienced derivation or AI derivation. The system concept results in the data relationships in Figure 3.1-4 Example System Data Flow below.

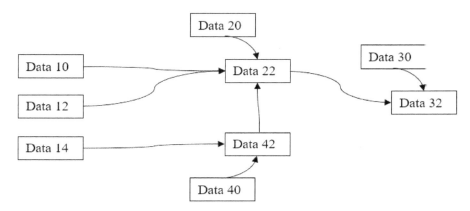

Figure 3.1-4 Example System Data Flow

The system development goal is to derive the output data: Data 32 based on the input data: Data 10, Data 12 and Data 14, and ensure that the output data: Data 32 meets the required timing.

To do which, the development needs to establish the $f1$, $f2$, $f3$ and the needed data, for example:

- to detect the objects based on the camera pixel signals, the development needs to establish the relationship between the sensed objects and the camera pixel signals.
- to output braking force based on the brake pedal angle, the development needs to establish the relationship between the brake motor output force and the brake pedal angle.

The relationship establishment between the output signals and the input signals belongs to the system feature development, not in the system engineering's scope System engineering is to provide the computing platform to run the $f1$, $f2$, $f3$ functions and store the needed data such as Data 40, Data 42, Data 20, Data 22, Data 30, Data 32, and make sure that functions can run correctly and reliably and the data can be stored and transferred correctly and reliably.

To make sure that the formula: Data 32 = $f3$ (Data 22, Data 30) can run correctly and reliably, which depends on:

- if the $f3$ operates correctly and reliably, which depends if the instructions that structure the function: $f3$ can be run by the Arithmetic Logic Unit (ALU), stored and transferred by the devices in the processor correctly and reliably, which in turn depends on the chip hardware and its functionality, for which the application can do little. What the system engineering needs to do is to design the suitable hardware for such operations.
- if the Data 22 and Data 30 are stored and transferred correctly and reliably, which will be done by the system engineering, such as error detection development, safety including the reliability development.

So, according to the system operation concept above, the system engineering can have the correct and complete information to:

- develop the required hardware and platform: From the functions: $f1$, $f2$, $f3$ and the data: Data 40, Data 42, Data 20, Data 22, Data 30, Data 32, the system engineering can design the hardware device needed and the communications between them.
- develop the required functionalities to realize the information.
- develop the required safety measures, such as error detection, reliability mechanisms to protect the data.

So, system engineering will have the defined, correct and complete development path based on the system operation concept. The data driven system engineering is to develop the system by following and focusing on the data path from the input signal to the output signal.

So, the system operation concept fully represents the required system using:

- the input data
- the middle data
- the output data
- the relationships between those data

It not only provides the information about how to derive each required output data, but also provides the information about the output data timing derivations.

3.1.2 Notation

This book introduces the following simple and straightforward notations consisting:

- Executable Procedure Node (EPN)
    - Executable Procedure Node Name and I/O Parameters
    - Data definition
    - Action
- Transitions between the EPNs

The syntax for the notation is the syntax of the programming language selected for the project, i.e., if the system under development is using C, then the EPN, data definition, the action instruction and the transition are defined using the select C programming language; if the C++ is selected, then the syntax is the C++; if java is selected, then the syntax is java, and so on.

The EPN structure is illustrated as in Figure 3.1-5 Executable Procedure Node (EPN) below, among which:

- The EPN structure is totally the same as a function definition in any programming language and an EPN must have the return value.

- The data definitions in an EPN are same as the data definition in any programming language, and the data can be defined as an EPN, as well, which is the same as in the programming languages that the functions can be treated as data. So that the definition is a recursive procedure to make it possible for automation
- Actions in an EPN operate the data to derive the results that are defined in the EPN.

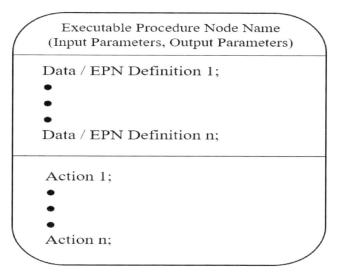

Figure 3.1-5 Executable Procedure Node (EPN)

The reasons why such notation is introduced are because:
- To make the notation design close to the required software program as much as possible
- To make the notation easy to use as much as possible

The final goal of development is to develop the source code that will run on the target system, for that, any activity that is not helpful to achieve the goal should be removed, any activities that have duplicated effect will be merged together to make it simple and easy, and for which, the author thinks the design notation should be simple and easy to use, so that the effects can be focused on the design, rather than the notations.

So that, the design using this notation will be very close to the final product, i.e., the source code for the system under development will simply be the collections of all the EPNs if all the EPNs are described in enough details using the selected syntax, in which, each EPN will form a function, the transitions between the EPNS will form the invoking of the functions.

The reason why using notations in a computing system design is because that the human letters and sentences are originated from the pictures, so the pictures or diagrams is more understandable comparing with the source code, it is much easier to express the design logic using notations.

As mentioned in the section of 3.1.1 Data Driven Development that the system development consists of:
- defining the input data and output data

- establishing the relationships between the input data and the output data

The first contents are the Requirement Engineering which will be described later.

The second contents are the system design, for which the notations are used to facilitate the design, and to achieve the system design, there are two middle targets:

- designing the functional components that make up the required system, which will be represented by the EPNs and the data definitions.
- designing the relationships between them, which will be represented by the actions in the EPNs.

The transitions between the EPNs are defined as the execution transmission conditions that transfers the computing system execution from one EPN to another. Note: it is the execution transmission, not the data transmission, the data transmissions are done by the EPNs.

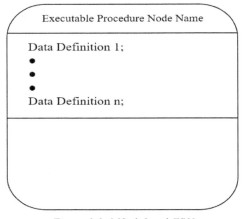

Figure 3.1-6 High Level EPN

The suggested approach to design system using such notation:

- At the beginning, use the comments instead of real programming sentences, and add more real programming sentences as the design progresses.
- At the high-level design, define the data only as illustrated in Figure 3.1-6 High Level EPN, because the data represent the functionalities, so if the data definitions are completed, then the system functionalities are fully described. And in this way, the system design can be clear, simple and easy to focus on the design logic, the important relationships in the system, and leave the detail design later.

3.1 Concept

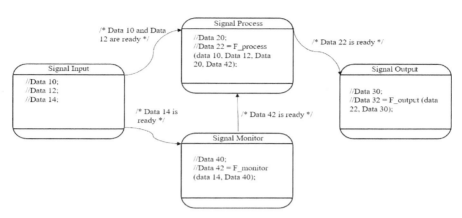

Figure 3.1-7 Example System

An example using such notation to describe a simple system is illustrated in Figure 3.1-7 Example System, in which all the transitions, data definitions and actions are comments.

As mentioned in the section of 3.1.1 Data Driven Development, the system design is to figure out the logic functions to derive the output signal from the input signals via some middle results.

The diagram in Figure 3.1-7 Example System above reveals the relationships between the defined data in Figure 3.1-8 Data Flow below, which clearly describes the data flow.

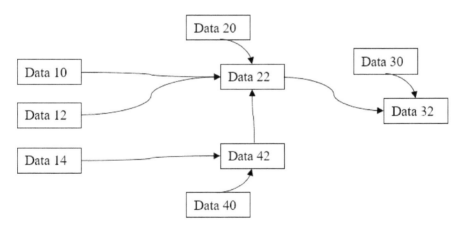

Figure 3.1-8 Data Flow

In which, the relationships between each data, especially the logic to derive the final output data: Data32 are clearly described, i.e., the logic about how each input signals derive the relevant middle results and how the middle results derive the output signal presents the relationship between those data, all that information describes how the system operates under the defined environment, so, the process of developing the information and descriptions above is the System Concept.

Then based on the information in the System Concept, the development can allocate each middle result on to the suitable devices, such as allocating the Data 42 on to the RPU, allocating the Data 22 on to APU, etc., figure out how those devices communicate each other, such as how the APU communicates with the RPU, how the Data 32 output to the outside, and the process of which is the System Structure Design.

 Based on the data relationship in Figure 3.1-8 Data Flow above, the system FMEA can be done in such way: Each data has either direct or indirect impact to the output data: Data 32, so, all failures in the system will impact the system function that is the Data 32. And furthermore, system safety development can follow those data relationship, as well. Since all data impact the system function, each data and each data transmission should implement the safe measures to achieve the required safety level.

 Those system development activities will be described in detail in the next sections.

3.1.3 Example

 To explain how to do the data driven system engineering activities, this book uses a Blind Spot Detection (BSD) as an example, such system consists of two object detection sub-systems: camera object detection system and the radar object detection system illustrated in Figure 3.1-9 BSD System below:

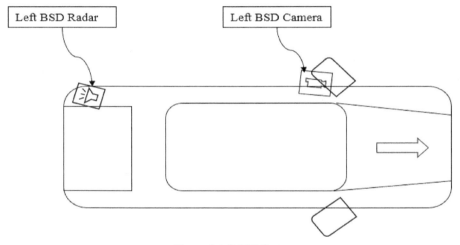

Figure 3.1-9 BSD System

 The camera sub-system and the radar sub-system are independent each other to detect the objects in the detection zone and output the detected objects to the electronic power steering (EPS) control ECU which will warn the driver and inhibit the vehicle left turn if there are the objects detected that may cause hazards.

 To simplify the example, it will only take the left side camera BSD sub-system into consideration, the data flow of which in the vehicle is illustrated in Figure 3.1-10 BSD Data Flow, which involves three ECUs:

- BSD Camera ECU, whose input signals are the optical signals from the environment, and the output signals are the camera pixel signals.
- BSD Calculation ECU that is the BSD ECU, which is the target ECU in discussion that calculates the objects based on the input camera pixels then output the detected objects.
- Power Steering Control, which controls the vehicle steering based on the input

signals from the BSD calculation ECU.

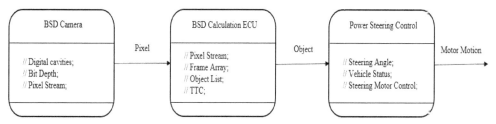

Figure 3.1-10 BSD Data Flow

The camera BSD system requirements are:

- Feature function: the camera BSD shall detect whether a vehicle is in or will be entering an area to the left side of the vehicle extending rearward from the outside left mirrors to a minimum approximately 5 meters beyond the bumper. This area is referred to as the detection zone illustrated in Figure 3.1-11 Left Blind Spot Detection below. The feature is designed to alert on targets entering the detection zone from the left, rear side, or front of the detection zone, and it is the input signal to the steering control module to prevent the left turn collision.

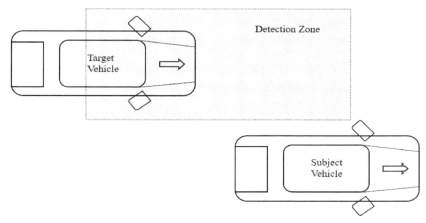

Figure 3.1-11 Left Blind Spot Detection

- The camera BSD shall send out the alert signal: CAN_BSD_Left_Alert (TTC) on the vehicle CAN bus with the periodic rate of 33 msec to represent if it detects the object vehicle(s) in the detection zone (TTC > 0), or no object vehicle (TTC = 0).
- If the Time to Collision (TTC) between the target vehicle and the subject vehicle is less than 2.5 seconds, the subject vehicle shall not allow the left turn.
- The left camera BSD system development shall be ASPICE compliant.
- The left camera BSD system shall be developed based on the AUTOSAR 4.3.0 or newer.
- The camera left Blind Spot Detection (BSD) system is a product of Automotive Safety integrity Level: ASIL B (D)
    (Note: the ASIL B (D) represents that the ASIL B is the result of decomposition from ASIL D, i.e., the left turn collision hazard is ASIL D that is prevented by

both Power Steering Control Module and both the camera BSD module and the radar BSD module, so the input signal from either the camera BSD or the radar BSD to the steering module is ASIL B (D))

- The BSD system shall be UN ECE 155 / 156 compliant and have the following cybersecurity protections:
    o The ECU programming contents shall be authenticated using RSA 4096, and the parameters should be encrypted using AES 128.
    o The access to the ECU contents shall be authenticated using AES 128, such as the disclose the ECU ID, modify the parameters, update the software.
    o The communication between the subject vehicle and the camera BSD system shall be prevented from unintended modification using AES 128.
- The left camera BSD shall use the camera sensor as the environment sensing device that is located at left rear comer of the rear bumper in the vehicle. The camera is managed by the BSD ECU including the power supply, intrinsic calibration and maintenance diagnostic services. The interface between the BSD ECU and the camera ECU are following:
    Physical Layer: Coax Connector
    Electrical Layer: LVDS
    Data Link Layer: TI FPD III
    Application Layer: YUV422 video output using the BT.656 digital video protocol
    1024 X 768, 30 fps.
- The vehicle communication bus interface is high speed CAN (500 kbits), and the message list is provided in the specification of BSD.DBC.
    Physical Layer: Two-wire, termination (SAE J2284-3)
    Data Link Layer: 11898-1, ISO 11898-2 and ISO11898-6.
- The camera BSD functions should be engaged only when Vehicle Ignition Status is either Run or Start, which is represented by CAN_Msg_Ignition that is a CAN signal with a periodic rate of 20 msec with the following values:
    o 0x0: Unknow
    o 0x1: Off
    o 0x2: Accessory
    o 0x4: Run
    o 0x8: Start
    o 0x15: Invalid
- The camera BSD functions should be engaged only when the vehicle speed absolute value is greater than 10 KM/H which is represented by CAN__Msg_Veh_Speed that is a CAN signal with a periodic rate of 20 msec with the range values: -100 KM / h ~ 200 KM / h.
- The ECU shall implement the diagnostic services defined in the Unified Diagnostic Service (UDS) according to the ISO 14229 standard with the DID, RID and DTC list specified in the BSD Diagnostic Specification, the functionalities of which shall include:
    o 3 defined diagnostic sessions:
        ▪ Default session
        ▪ Programming session
        ▪ Extended session

- ○ Session Control Services: $10
- ○ Data maintenance services including
  - ■ Read Data: $22
  - ■ Write Data: $2E
- ○ DTC services:
  - ■ Read DTC: $19
  - ■ Erase DTC: $14
  - ■ Control Setting: $85
- ○ Routine services: $31
- ○ Security services: $27
- ○ ECU Reset services: $11
- ○ Programming services:
  - ■ Communication control: $28
  - ■ Request Download: $34
  - ■ Transfer Data: $36
  - ■ Request Exit Transfer: $37
- ○ Tester Present Services: $3E
- The camera BSD system is power by the vehicle power and shall work in the range from 8 V to 16 V, when the power is out range, the camera BSD shall send out the error status on the vehicle CAN bus.
- The camera BSD system shall work in the temperature range from -40°C ~ +80°C, when the temperature is out range, the camera BSD shall send out the error status on the vehicle CAN bus.

In the camera BSD system realization, the ECU will use a microcontroller with the following features:
- There are two RPU core running in the "Lock-Step" mode for the safety tasks.
- There are two APU cores for the high-performance tasks.
- One GPU with the Pixel processor
- The FPGA provide the control logic for audio interface, which is the redundant output warning signal in this application.
- The SHE unit provide the crypto algorithm engine for the cybersecurity functions.
- Internal RAM and EEPROM provide the memory for the middle result storage.

Note:
- This BSD system has the ASIL D features, which is chosen for the purpose to describe the safety handling such as safety requirement decomposition, and to point out that such ASIL D requirement decomposition to ASIL B(D) requires that the hardware safety requirement remains at the ASIL D, which may be the issue to some real applications in the market.
- This camera BSD is only an example, not intended to be real control module in any aspect that is much more complicated than this example, which should have more functionalities' requirements, such as diagnostic services, fail-safe measures, features and parameter configurations, which are not fully covered by the example, Without the full requirements coverage, it would cost serious issues for the real products in both quality and safety.
- In rest of the book, if there is not the specific indication, the BSD system means the

3.1 Concept

left side camera BSD sub-system.

3.2 Requirement Elicitation

The first step in every development is the Requirement Elicitation.

The goal of requirement elicitation is to collect all the requirements that fully describe the system under development consisting of:
- Feature function
- Maintenance function represented by the diagnostic services
- Input and output interfaces and signals
- Safety
- Cybersecurity
- Quality Management

The ECU requirements are divided into two types:
- Technical Requirements (sometimes they are called: functional requirements)
  - Computable
  - Non-computable
- Non-Technical Requirements (sometimes they are called: non-functional requirements)

Among above, the technical requirements are the main target of system development, and this book introduces the "two steps" approach to do the technical requirement elicitation that will be described later in this section.

The technical requirements are the following:
- The information that describes the system's external behaviors, which are the system output signals.
- The information that will trigger the changes of system's external behaviors, which are the system input signals.
- The information about the system external behaviors' constraints, which are the system performance requirement.

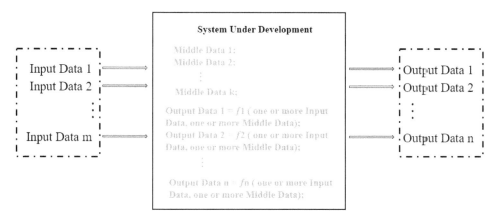

Figure 3.2-1 Technical System Requirement

65

3.2 Requirement Elicitation

From the system black box point of view, the Technical Requirements are the information that technically describe the required product illustrated in the dash boxes of Figure 3.2-1 Technical System Requirement, which are the information related to the input and output signals and divided into two types:

- Computable requirement: which are readable to a computer.
- Non-computable requirement: which are not readable to a computer.

If the requirements are computable, then they can be processed by the computer that are used in the product or system under development. So, they must be either input signals or output signals.

If the requirements are not computable, then they must be testable, i.e., they can be expressed in some test cases. The examples of non-computable requirements are performance or quality requirements.

The computable requirements are testable from their nature. So, all the technical requirements must be testable.

If any technical requirements are not testable, then those requirements are not valid, which need to be modified to be testable.

The Non-Technical Requirements are those for the development management or processes, which are mainly targeted at the development quality. Those aspects of system requirements are always provided using the organization regulations and the industrial standards.

The reason why the requirements are provided using the industrial standards are because the contents of standards are well known, and it is efficient and accurate to provide the requirements in such way. How to handle and implement those requirements will depend on the customers and suppliers' processes. If the organizations have the processes that will engage the quality control requirements that are required by the standards, then the requirements can be elicited in such way:

- The system development procedures shall be ASPICE compliant.
- The system shall be developed by following ISO 26262 and reach the ASIL B.
- The system shall be developed by following the UN ECE 155 / 156 and use the RSA 4096 for the programming authentication, AES 128 for communication authentication.
- The system development shall be based on the AUTOSAR release 4.0.

Among the examples above, there are all mainly and commonly used regulations and standards in the automotive ECU development, the ASPICE and ISO2626 are two commonly required standards to cover all aspects of product development even including the managements of suppliers and production, the UN ECE 155 / 156 regulations are specific for the cybersecurity and the AUTOSAR is for the ECU platform that can be seen as the operation system in the automotive ECUs.

It is almost the case for all automotive OEMs and suppliers to use the regulations and industrial standards to specify certain system requirements using regulations and standards. However, if the organizations' processes don't cover the development quality control measures that are required by the standards, then the requirements need to be extracted specially for the required product.

In some cases, regulations and standards provide not only the quality control measure requirements, but also the technical requirements including the requirements for the internal structures, in those cases, it is better to extract the required technical requirements explicitly from the regulation and standards, for example:

- The requirement: "The product shall be compliant with ASIL B" can be translated as:
  - The product hardware Probabilistic Metrics for Hardware Failures rate shall not be more than 100FIT,
  - The product hardware Single-Point Fault Metric shall not be less than 90%,
  - The product hardware Latent Fault Metric shall not be less than 60%.
- The requirement: "The system shall be developed by following the UN ECE 155 / 156 and use the RSA 4096 for the programming authentication, AES 128 for communication authentication" can be translated as:
  - The product content programming shall be authenticated using the RSA 4096 algorithm.
  - All specified authenticated diagnostic services shall be securely guarded using AES 128.
  - All specified authenticated CAN messages shall be encrypted and decrypted using AES 128.

In North America automotive industry, all the big three have their specific way to document the required product:

- For GM, it is the Component Technical Requirement Specification (CTRS)
- For Chrysler, it is the Product Function (PF) and Vehicle Function (VF) documents
- For Ford, it is the ECU specific Functional Specification

So, for the projects from those three OEMs, those documents are the keys for the requirement elicitation.

During the requirement elicitation phase, it is worth to point out that in most cases, the customers often provide some logic relationship between the signals as the "requirements", which are actually the outcomes from design. Ideally, the requirements should be given in such way that the system under development is like purchasing a product from Off-The-Shelf, i.e., the requirements should be only about the interfaces and the functionalities to describe what are needed from the outsides, and the design is the developers job. However, in the reality, a quite some automotive ECU requirements are given from the designer's point of view, which becomes the constraints to the development.

The benefit of clearly distinguish between the system requirements and system design is to clearly define the development boundary and scope, so that the development responsibility is clearly allocated to the dedicated teams to speed up the development.

A systematic way to do the requirement elicitation is to use the Two-Steps System Requirement Elicitation approach:

- First Step: elicit the physical signals that are input to and output from the required system.
- Second Step: categorize and detail the functional requirements according to the physical signals from the first step.

The first step is to follow the Figure 2.2-2 ECU Input and Output Signals which illustrates a typical automotive ECU and related I/O signals:

- Vehicle Communication, which is mainly based on the serial communication signals and divided into two parts: signal interface and content
  - Signal Interface: automotive customers always have the specific requirements about the communication interfaces to ensure that the ECUs under development can communicate with others in the vehicle. So, the requirements about the communication bus type (CAN, Ethernet, LVDS, FlexRay, etc.), the communication transmission rate: (10 Gbps, 500 Mbps, 200 Mbps, etc.) and which ECUs will be communicated with by the subject ECU should be fully clarified.

    All automotive vehicle communications can be described accurately using the industrial standards to give the physical layer requirements, electronic layer requirements, data link layer requirements and application layer requirements. For example: the CAN physical and electronical layer requirements shall follow the ISO 11898-2, which means that the CAN is the high-speed CAN (bit speeds up to 1 Mbit/s on CAN, 5 Mbit/s on CAN-FD) and uses a linear bus terminated at each end with 120 Ω resistors.
  - Signal Content: which is mainly for:
    - the status communication, i.e., the system acquires the vehicle status and reports its own status to the vehicle.
    - The maintenance services requests
- Environment Signal: Those are the analog and digital signals including
  - The vehicle environment sensing signals, such as the object detection if the ECU is the one for camera or radar sensor.
  - The ECU environment temperature signals.
  - Any other input signals, such as the ECU installation position, fuel level signals, door ajar switch signals.
  - Any other output signals, such as PWM to drive motors, LEDs, camera heating, etc.
- NVM: The information stored in the NVM needs to follow certain requirements, such as format, region, security.
- Power: How the power from the vehicle connects with the ECU is important requirement, which is either permanently connected all the time, or switched on and off by the vehicle ignition switch. The requirement impacts the ECU sleep and wakeup strategy.

The second step is to categorize and detail the functional requirements that are embedded in the physical signals. The automotive ECU functionalities are categorized in 3.4.2 System Structure Design, which are:

- Feature Function
- Diagnostic Service
- Cybersecurity Function
- Serial Signal Manager
- Application Mode Manager

All those functionalities are based on the physical input and output signals that elicited from the first step. For the BSD example in 3.1.3 Example, the functionalities can be analyzed as below:

- Communication:
    - CAN vehicle bus:
        - Feature functions: which need the vehicle ignition status and vehicle speed signals from the bus, and output the required object detection including the TTC to the bus.
        - Diagnostic services: all of the diagnostic service requests and responses are based on the bus.
        - Cybersecurity functions: which require the support from the diagnostic services and serial signal manager.
        - Serial Signal Manager: which is the dedicated component for this CAN vehicle bus to handle the serial communication.
        - Application Mode Manager: which needs the diagnostic services and the serial signal manager support to report the ECU status to the vehicle, such as the ECU is in the initialization status or fault status.
    - LVDS:
        - Feature functions: which calculate the detected objects based on the camera pixel steam input signal.
        - Diagnostic services: which are for the camera ECU maintenance that are transferred from the diagnostic client to the camera via the BSD ECU.
- Environment:
    - Application Mode Manager: which needs the ECU working temperature signals.
- NVM:
    - Diagnostic services: which provide the programming services to the NVM contents.
    - Application Mode Manager: which will monitor the interactions between the NVM manager in the AUTOSAR and the NVM devices, and will monitor the secure booting activities as the NVM content authentication is required.
    - Cybersecurity functions: which will securely guard the NVM contents as the authentication is required.
- Power:
    - Application Mode Manager: which will handle the sleep and wake requests, and monitor the power supply voltage to ensure the stable working power supply.

The reasons and benefits of using the two steps approach to elicit the system requirements are:

- The first step can be done easily and quickly, which is just to follow the Figure 2.2-2 ECU Input and Output Signals, and from which:
    - All the signals that the ECU needs to handle are known, which provide the complete coverage of the ECU capability, in another words, if there is functionality that does not interact with the ECU, then it won't belong to

the ECU because the ECU cannot handle it.

- o All the input and output devices needed are known, which can provide valuable information to the quotation or to estimate the project plan.
- o The categories of input and output signals provide valuable information for the system verification which needs to know what signals from what devices interact with the ECU.

- The second step can provide the detailed and categorized functionalities based on the input and output signals from the first step, and from which:
    - o What functionalities needed are known, and which are categorized based on the functional components that will be implemented in the 3.4.2 System Structure Design.
    - o What functionalities need what input or output devices are known, which are helpful to the system design and system verifications.

Among above, from the requirement point of view, the first step covers the "Width" and the second step goes into the "Depth".

During the requirement elicitation, the requirements collected should be managed using the requirement management tool. At the top level from the customer side, usually the requirements illustrated in Figure 2.2-2 ECU Input and Output Signals are given in the text format, so collecting all the requirement items in a list is easy to read, and the tools to manage them are, for example, the Microsoft Word, Excel, IBM DOORS, which can organize the requirement items in chapters, sections and paragraphs, etc., the format of which is like this book.

If the requirements are given using some diagrams, such as Hatley-Pirbhai method or SysML diagrams, then one or a few text sentences can be created to express them, such as: the functional logic shall follow the diagram below, …, then use those items as the leading requirements in the requirement list to make them more readable.

In the SysML, there are two diagrams to express the requirements: Requirement Diagram and Package Definition Diagram, which, comparing with the table requirement management tool above, are neither easy to use nor to read. So, from this book point of view, the better way to manage the requirements is to use list alike tools, rather than the diagrams.

In the BSD example, the requirement elicitation will be exact same as in the section of 3.1.3 Example.

Check List 1 - Requirement Elicitation

- Do the elicited requirements fully describe the required system?
- Are the system functionality and performance requirements defined?
- Are the system safety requirements defined?
- Are the system cybersecurity requirements defined?
- Are the system diagnostic requirements defined?
- Are the system AUTOSAR requirements defined?
- Is the system communication including the sleep and wakeup requirements defined?
- Is the system power including the power on and off requirements defined?
- Is the system environment (temperature, voltage, object sensing, interacting with the host vehicle) requirements defined?
- Are the system external memory (programming, authentication and confidentiality) requirements defined?
- Are the system development management requirements defined?

3.2 Requirement Elicitation

3.3   Requirement Engineering

The goal of requirement engineering is to analyze, translate and categorize the collected requirements in the requirement elicitation phase to the suitable categories to be managed; and to the suitable format to be executed.

For the computable requirements among which, the requirement engineering needs to:

- Transfer the computable technical requirements to be either the input or output signals requirements.
- Transfer the non-computable technical requirements to be the test cases, which should be either the performance or quality test cases.

In addition to the technical above, the quality requirements need to be analyzed, as well, which are always specified using the regulations and industrial standards consisting two aspects:

- Development activities, for example, in the system development phase of ISO 26262, the FMEA activity is suggested; in the ASPICE, the software detailed design is required.
- In some cases, the performance quantitative criteria are provided by the requirements, such as, in the Part 5 of ISO 26262, the product hardware Probabilistic Metrics for Hardware Failures rate, Single-Point Fault Metric and the Latent Fault Metric should meet the dedicated ratings.

In the ASPICE, the purpose of system requirement analysis is to transform the defined stakeholder requirements into a set of system requirements that will guide the system design. And in the conventional requirement engineering, the goal of requirement engineering is to decompose the system requirements into the system architecture design input specifications.

Both of above are vague, and the activities in the Requirement Engineering cannot be measured accurately.

This book makes the goal and scope of requirement engineering in the computing system development specific, accurate and measurable:

- Goal: the requirement engineering is to prepare the input and output signal that will be used by the formulas of system operation concept described in 3.1.1 Data Driven Development.
- Scope: the requirement engineering is to use the computer executable information to describe the system under development illustrated in Figure 1.1-1 Computing System, which consists only two types of information: Signal and Test Case.
- Measurement:
  - o Signals, either input or output signals, shall be computer readable, for which, the following information should be available:
    - Signal Name
    - Signal direction (Input or Output)
    - Signal Length (bit)
    - Value Range (min – max)
    - Unit
    - Accuracy (%)
    - Frequency
  - o Test cases shall be executable to the system under development, which

should have the information:
- The signal(s) to be tested
- The configuration(s) of the test(s)
- The criteria for the test(s) to pass or fail

To achieve the goal of transferring the computable requirements to be executable by the computer, for the signal requirements, the computer variables are added into the requirements to represent the requirement signals; for the test case requirements, the test measurements that represented by the computer variables as well need to be described.

Once the signals become computer executable, then they can be either put into the relationship formula given in the system operation concept to do the system architecture design, or executed to measure the target signals.

It is worth to point out that this development phase is called: "Requirement Engineering", and any engineering activity must have two aspects:
- Analysis
- Design

The common mistakes that happen very often are that:
- The requirement engineering activity is done literally, i.e., the sentences in the requirement documents will be literally broken into simple sentences or sub-paragraphs without any further analysis or design.
- The requirement engineering activity is done by the "Requirement Engineers" who are in most cases not familiar with the system design or product design.

The consequences from such requirement engineering are:
- The requirement specifications are huge.
- The requirement specifications hardly provide any optimized or value-added ideas to the next development that is the system architecture design, so that in most cases, the system architects need to re-do the activities: analyzing the requirement specifications, figure out the logics among them, which is a huge waste.
- In some worse cases, the requirement specifications are transferred from the customers, to the requirement engineering phase, to the system architecture design phase, to the software requirement engineering phase, to the software architecture design phase, then down to the software detailed design phase without any analysis or design except mainly that the sentences are broken down.

The keys to successful requirement engineering are:
- Figure out the relationship between the technical requirements, especially the relationships to the output signals. There is not any input or output signal that is not related to someone else in a computing system.
- Prepare and optimize the information that is needed by the system architecture design, such as logically combine or merge some information together to optimize the signals; decompose some complex information to simplify the relationships.

Taking the BSD system that is introduced in the section of 3.1.3 Example as an example, the requirement engineering is done as following:

For the requirement: "the camera BSD shall detect whether a vehicle is in or will be entering an area to the left side of the vehicle extending rearward from the outside left mirrors to a minimum approximately 5 meters beyond the bumper. This area is referred to as the detection zone. The feature is designed to alert on targets entering the detection zone from the left, rear side, or front of the detection zone, and it is the input signal to the steering control module to prevent the left turn collision."

This is neither an input nor an output signal requirement, rather, is a system performance requirement, so, which can be described as the test cases below:

Test Configuration:

Set the subject vehicle at the forward speed as 50 KM/h on the straight even road.

Test Case 1: Set the target vehicle at the forward speed as 50 + X1 KM/h to approach the subject vehicle from the left adjacent line, the expected output signal: CAN_BSD_Left_Alert (TTC) from the BSD module should be Y1 when the target vehicle's distance is Z1 meters from the subject vehicle's rear bumper.

Test Case 2: Set the target vehicle at the forward speed as 50 + X2 KM/h to approach the subject vehicle from the left cross line to the adjacent line when the target vehicle' head is at W1 meters from the subject vehicle's rear bumper, the expected output signal: CAN_BSD_Left_Alert (TTC) from the BSD module should be Y2 when the target vehicle's distance is Z2 meters from the subject vehicle's rear bumper.

Test Case 3: Set the target vehicle at the forward speed as 50 − X3 KM/h to approach the subject vehicle from the left adjacent line, the expected output signal: CAN_BSD_Left_Alert (TTC) from the BSD module should be Y3 when the target vehicle's distance is Z3 meters from the subject vehicle's rear bumper.

Among the test cases above, the parameter: X1, X2, X3 are the testing variables which can be changed during the test to the multiple object detections at the different speed.

For the requirement: "The BSD shall send out the alert signal: CAN_BSD_Left_Alert (TTC) on the vehicle CAN bus with the periodic rate of 33 msec to represent if it detects the object vehicle(s) in the detection zone (TTC > 0), or no object vehicle (TTC = 0)."

This is an output signal requirement which is already a computer readable requirement which can be described as:

The system shall calculate and send out the CAN signal: CAN_BSD_Left_Alert (TTC) = f_time (Left_Video_LVDS) at the periodic rate of 33 msec, among which, the output signal transmission rate: 33 msec is very important requirement, which is the system performance requirement, i.e., the output signal update rate (33 msec) needs that all the calculations to derive the signal need to be done within such 33 msec periodic time ideally, however, such performance requirement is not feasible that will be explained in the section of 3.4.4.3 Latency, then this periodic time means that the output signal shall be updated "as soon as possible" if it cannot be done within the required time.

For the requirement: "If the Time to Collision (TTC) between the target vehicle and the subject vehicle is less than 2.5 seconds, the subject vehicle shall not allow the left turn."

This is a safety requirement, and specifically, it is ASIL D requirement for the whole left BDS system including the left camera, left BSD and the electronic power steering (EPS) control module.

From this requirement, it explains the critical timing requirements to each ECU in the BSD system though the explicit requirement is not given here.

The TTC (2.5 seconds) is the time that the subject vehicle starts the left turn, which is the T1 illustrated in Figure 3.3-1 Detection Time, however, the target vehicle enters the detection zone at T0, and the whole BSD system will take the time = T1 - T0 to transfer the signal and take the action at the electronic power steering (EPS) control module. So, the detection time in the worst case for the BSD system to detect the target vehicle is: TTC (2.5 seconds) + (T1-T0), among which, the (T1-T0) is the time for the whole BSD system to operate that is started from the camera detect the target vehicle to the electronic power steering (EPS) control module to act.

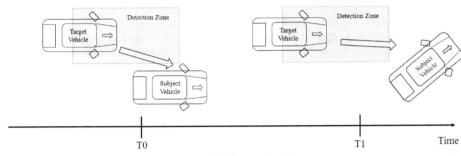

Figure 3.3-1 Detection Time

This requirement does not explicitly have the requirement for the left camera BSD system, and as to the whole BSD system timing performance, which will be described in 3.4.4.3 Latency. So, this requirement will be taken as an explanation for the left camera BSD system.

For the requirement: "The left camera BSD system development shall be ASPICE compliant."

This is the development process requirement that is applicable for every automotive ECU development, which requires that the development organization should establish the development processes that are described in 2.2.2.1 ASPICE.

For the requirement: "The left camera BSD system shall be developed based on the AUTOSAR 4.3.0 or newer."

This is the technical requirement that is applicable for most automotive ECU development, which requires that the required ECU platform should be constructed based on the AUTOSAR illustrated in 2.2.2.2 AUTOSAR. There are the dedicated AUTOSAR supplier who provide the software packages, and the ECU suppliers are only the integrators regarding to this point, how to integrate the AUTOSAR into the required ECU will be described in 3.4.2 System Structure Design.

So, this requirement can be described as: The left camera BSD system shall be integrated into the AUTOSAR platform released 4.3.0 or newer.

For the requirement: "The camera left Blind Spot Detection (BSD) system is a product of Automotive Safety integrity Level: ASIL B (D)."

This is a requirement covering both non-technical and technical requirements.

For the non-technical requirements, it requires the development to follow required processes including all aspects of automotive ECU development. The detailed requirements and how to satisfy ISO 26262 will be described in 3.4.6.4 ISO 26262 Compliance.

The requirement ASIL B (D) means:
- The object detection software is ASIL B
- The hardware that supports the object detection software must be ASILD, however, for most camera ECUs and BSD ECUs, it is not feasible to reach ASIL D, so in this example, the object detection hardware requirement is still set to ASIL B, and it is the implementation for all the object detection system in the current market. The reason why this example uses the ASIL B (D) rating because it is the common requirement and this book intends to point out the potential issue in 3.4.6.3 BSD Safety.

The only measurable technical requirements from ISO 26262 ASIL B rating are following:
- The product hardware Probabilistic Metrics for Hardware Failures rate shall not be more than 100FIT,
- The product hardware Single-Point Fault Metric shall not be less than 90%,
- The product hardware Latent Fault Metric shall not be less than 60%.

For the requirement: "The BSD system shall be UN ECE 155 / 156 compliant and have the following cybersecurity protections:

The ECU contents shall be authenticated using RSA 4096, and the parameters should be encrypted using AES 128.

The access to the ECU contents shall be authenticated using AES 128, such as the disclose the ECU ID, modify the parameters, update the software.

The communication between the subject vehicle and the camera BSD system shall be prevented from unintended modification using AES 128."

First, the two regulations are below:
- UN Regulation No. 155 – Cybersecurity
- UN Regulation No. 156 - Software Updates

Similar to ISO 26262, those two regulations cover not only the technical but also the non-technical requirements, the No. 155 covers all aspects of development related to cybersecurity, and the No. 156 is specifically about the software updates.

The system requires the RSA 4096 for the asymmetry encryption and decryption, and AES 128 for the symmetry encryption and decryption. The cryptographic algorithms and used key length are important information which will be explained in the section of "The system cybersecurity requirements" of 3.4.3.1 Microcontroller Selection.

To satisfy the requirement of "The ECU programming contents shall be authenticated using RSA 4096, and the parameters should be encrypted using AES 128". as the common implementation, the programming of an automotive ECU is done using the dedicated software component: Bootloader, so this requirement can be described as the following bootloader functionality requirements:
- The ECU bootloader shall verify the access authentication using the AES 128 security algorithm which is called as ECU locking / unlocking functions.
- The ECU bootloader shall verify the programming contents' signature as the authentication using the RSA 4096 security algorithm.

- The ECU bootloader shall encrypt the programming parameter contents stored in the external flash memory using the AES 128 security algorithm.

To satisfy the requirement of "The access to the ECU contents shall be authenticated using AES 128, such as the disclose the ECU ID, modify the parameters, update the software", the following implementations need to be done:

- The access to the ECU testing interfaces shall be authenticated using AES 128.
- The access to the critical information inside of the ECU shall be authenticated using AES 128.
- The executions of ECU diagnostic routines shall be authenticated using AES 128.

To satisfy the requirement of "The communication between the subject vehicle and the BSD system shall be prevented from unintended modification using AES 128.", which will be done:

- The message values shall be authenticated, which will be protected using the Protect Value method that encrypt the messages' values from the senders and decrypted by the receivers using AES 128.
- The messages shall be prevented from the playback attack that will copy and insert the messages again to manipulate the communication. The protection uses the message counter, i.e., each message will have the message counter to indicate the current message sequence number, which can be verified by the receiver to make sure the message is not either duplicated or missed.

Those cybersecurity requirements are still not detailed enough to be computer readable, because each of them still can be decompose into many sub-system requirements, and in the real development, there are some details that should be clarified, such as which message authentication algorithm should be used, how the signatures are located, which AES 128 key should be used to securely lock and unlock the diagnostic and test interfaces, such as the JTAG, XCP. However, the cybersecurity requirements are the protection against the external threats and attacks, they don't impact the system output signals, so, the detailed requirements will be described in the dedicated section in the book.

For the requirement: "The left camera BSD shall use the camera sensor as the environment sensing device that is located at left rear corner of the rear bumper in the vehicle. The camera is managed by the BSD ECU including the power supply, intrinsic calibration and maintenance diagnostic services. The interface between the BSD ECU and the camera ECU are following:

Physical Layer: Coax Connector
Electrical Layer: LVDS
Data Link Layer: TI FPD III
Application Layer: YUV422 video output using the BT.656 digital video protocol
1024 X 768, 30 fps."

In a vehicle, the video signals transmission between the ECU and the sensors or HMI devices uses commonly the Low-voltage differential signaling, or LVDS standard, so this requirement item can be described as:

The system shall read the camera video signals from the LVDS port into the variable: Left_Video_LVDS, which is an input camera pixel signal with the value of 1024 X 768, the signal speed is: 30 fps.

The system shall read the camera input pixel signals from the LVDS

The system shall provide and manage the power supply to the camera.

The system shall provide and manage the intrinsic calibration to the camera.

The system shall provide and manage the gateway functions for the camera s diagnostic services.

For the requirement: "The vehicle communication bus interface is high speed CAN (500 kbits), and the message list is provided in the specification of BSD.DBC.
Physical Layer: Two-wire, termination (SAE J2284-3)
Data Link Layer: 11898-1, ISO 11898-2 and ISO11898-6."

This is a CAN bus interface requirement, which is for the hardware CAN transceiver selection. In the current market, there are many electronic elements that meet such requirement, and the automotive OEMs always provide the list of approved CAN transceivers for the supplier to choose.

Regarding to the communication contents, the automotive OEMs always provide the message list using either the MS Excel sheet or Vector DBC file that contains all the messages between the ECU and the host vehicle, in this way, the ECU will know what and when it should receive and transmit, which will enhance the vehicle network efficiency, safe and secure. So, this requirement can be described as:

The vehicle CAN interface shall be compliant with the following standards:
Physical Layer: Two-wire, termination (SAE J2284-3)
Data Link Layer: 11898-1, ISO 11898-2 and ISO11898-6

The communication contents shall follow the BSD.DBC specification.

For the requirement: "The camera BSD functions should be engaged only when Vehicle Ignition Status is either Run or Start, which is represented by CAN_Msg_Ignition that is a CAN signal with a periodic rate of 20 msec with the following values:

- 0x0: Unknow
- 0x1: Off
- 0x2: Accessory
- 0x4: Run
- 0x8: Start
- 0x15: Invalid"

This is a typical automotive ECU feature function enabling requirement based on the typical vehicle ignition status signal, which can be described as:

- The system shall read the CAN signal: CAN_Msg_Ignition to determine the vehicle ignition status.
- The system shall only enable the BSD when the vehicle ignition status is either Run or Start.
- The signal periodic time is 20 msec.

For the requirement: "The camera BSD functions should be engaged only when the vehicle speed absolute value is greater than 10 KM/H which is represented by CAN_Msg_Veh_Speed that is a CAN signal with a periodic rate of 20 msec with the range values: -100 KM / h ~ 200 KM / h."

This is another typical automotive ECU feature function enabling requirement based on

the typical vehicle speed signal which can be described as:
- The system shall read the CAN signal: CAN_Msg_Veh_Speed to determine the vehicle speed whose value range is -100 KM / h ~ 200 KM / h.
- The system shall only enable the BSD when the vehicle speed absolute value is more than 10 KM / h.
- The signal periodic time is 20 msec.

For the requirement: "The ECU shall implement the diagnostic services defined in the Unified Diagnostic Service (UDS) according to the ISO 14229 standard with the DID, RID and DTC list specified in the BSD Diagnostic Specification, the functionalities of which shall include:
- 3 defined diagnostic sessions:
  - Default session
  - Programming session
  - Extended session
- Session Control Services: $10
- Data maintenance services including
  - Read Data: $22
  - Write Data: $2E
- DTC services:
  - Read DTC: $19
  - Erase DTC: $14
  - Control Setting: $85
- Routine services: $31
- Security services: $27
- ECU Reset services: $11
- Programming services:
  - Communication control: $28
  - Request Download: $34
  - Transfer Data: $36
  - Request Exit Transfer: $37
- Tester Present Services: $3E"

This is the typical automotive ECU diagnostic service requirements, all the automotive OEMs have the same requirements with the project specific Data Identifier (DID), Routine Identifier (RID), Diagnostic Trouble Code (DTC) implementations.

In most projects, the OEMs will require that the diagnostic service functionalities should be bought from the dedicated software suppliers, the ECU suppliers are only the integrators, so, this requirement can be described as:

The left camera BSD system shall implement the DID, RID and DTC list specified in the BSD Diagnostic specification.

For the requirement: "The camera BSD system is power by the vehicle power and shall work in the range from 8 V to 16 V, when the power is out range, the camera BSD shall send out the error status on the vehicle CAN bus."

This is the typical automotive ECU enabling requirement based on the vehicle power supply voltage signal, which can be described as:
- The BSD system working power voltage is the range: 8V <= Working Voltage <=

16 V.

- The BSD shall be disabled and send out the warning signal on the vehicle CAN bus if the BSD input power is out of range.

For the requirement: "The camera BSD system shall work in the temperature range from -40°C ~ +80°C, when the temperature is out range, the camera BSD shall send out the error status on the vehicle CAN bus."

This is another typical automotive ECU enabling requirement based on the ECU environment temperature signal, which can be described as:

- The BSD system working temperature range shall be: -40°C <= T <= +80°C.
- The BSD shall be disabled and send out the warning signal on the vehicle CAN bus if the BSD environment temperature is out of range.

After the requirement engineering is done, the requirements should have the following sections:

System Description:

Every specification should have the description, which describes what the contents are about as clear as possible, but not as detailed as possible. The purpose is to make the readers understand and get into the contents easily, so that:

- The specification can be shared easily in the development.
- The specification can be re-used easily later.

System Description
The specification is for the Blind Spot Detection (BSD) system, which is one of sub-system of the vehicle left camera BSD system consisting of the left BSD camera, this BSD system and the Electronic Power Steering (EPS) control module illustrated in the Figure 3.1-10. The BSD system detects the objects in the detection zone based on the camera input and outputs the detected objects to the power steering control ECU: - Left BSD Camera ECU, whose input signals are the optical signals from the environment, and the output signals are the camera pixel signals. - Left BSD ECU, which is the target ECU in development that calculates the objects based on the input camera pixels then outputs the detected objects. - Electronic Power Steering (EPS) ECU, which controls the vehicle steering based on the input signals from the BSD ECU. The left camera BSD shall detect whether a vehicle is in or will be entering an area to the left side of the vehicle extending rearward from the outside left mirrors to a minimum approximately 5 meters beyond the bumper. This area is referred to as the detection zone illustrated in Figure 3.1-11 Left Blind Spot Detection. The feature is designed to alert on targets entering the detection zone from the left, rear side, or front of the detection zone, and it is the input signal to the electronic steering (EPS) control module, if the Time to Collision (TTC) between the target vehicle and the subject vehicle is less than 2.5 seconds, the subject vehicle shall not allow the left turn.

Table 3.3-1 BSD System Description

Quite common, some specifications do not have the specification description, only have the scope of applicability at the beginning, which make it difficult for the readers to understand, especially for the readers who are familiar with the contents.

Input & Output Signal:

The input and output signal section contains all the signals that the system will handle, each signal should have the attributes of name, direction, signal length and range, unit, accuracy and frequency.

The combination of Input & Output signal and the system description cover the feature function, which provides the detailed information for the next development.

Input & Output Signal						
Signal Name	**Signal Direction (I/O)**	**Signal Length (bit)**	**Value Range (min / max)**	**Unit**	**Accuracy (%)**	**Frequency**
CAN_BSD_Left_Alert (TTC)	O	16	0 / 65535	ms	95	33 ms
Left_Video_LVDS	I	1024X768	NA	Pixel	NA	33 ms
CAN_Msg_Ignition	I	8	0 / 0x15	NA	100	20 ms
CAN_Msg_Veh_Speed	I	16	-100 / 200	KM/h	98	20 ms
ADC_Env_Temperature	I	8	-40 / 80	C	90	continuous
ADC_Veh_Power	I	8	0 / 24	V	90	continuous
CAN_Diag_Request	I	8	NA	NA	NA	spontaneous
CAN_Diag_Response	O	8	NA	NA	NA	spontaneous

Table 3.3-2 BSD Input & Output Signal

Functionality Requirement:

All the BSD functionality requirements except the feature function are listed here.

In theory, all the functionalities can be described using the input and output signals, however, as discussed at the beginning of this section, sometimes it is more efficient and clearer to describe some functionality requirements using the nature languages.

Functionality Requirement
Safety
The hardware devices that support the object detection software is ASIL B.
The object detection software is ASIL B.
The product hardware Probabilistic Metrics for Hardware Failures rate shall not be more than 100FIT
The product hardware Single-Point Fault Metric shall not be less than 90%.
The product hardware Latent Fault Metric shall not be less than 60%.

Cybersecurity Requirement
The ECU bootloader shall verify the access authentication using the AES 128 security algorithm which is called as ECU locking / unlocking functions
The ECU bootloader shall verify the programming contents' signature as the authentication using the RSA 4096 security algorithm
The ECU bootloader shall encrypt the programming parameter contents stored in the external flash memory using the AES 128 security algorithm
The access to the ECU testing interfaces shall be authenticated using AES 128
The access to the critical information inside of the ECU shall be authenticated using AES 128.
The executions of ECU diagnostic routines shall be authenticated using AES 128
The message values shall be authenticated, which will be protected using the Protect Value method that encrypt the messages' values from the senders and decrypted by the receivers using AES 128
The messages shall be prevented from the playback attack that will copy and insert the messages again to manipulate the communication. The protection uses the message counter, i.e., each message will have the message counter to indicate the current message sequence number, which can be verified by the receiver to make sure the message is not either duplicated or missed

AUTOSAR
The left camera BSD system shall be integrated into the AUTOSAR platform released 4.3.0 or newer

Diagnostic Service
The left camera BSD system shall implement the DID, RID and DTC list specified in the BSD Diagnostic specification.
The left camera BSD system shall implement the diagnostic service gateway to the camera

Communication Interface
Camera Interface: Physical Layer: Coax Connector Electrical Layer: LVDS Data Link Layer: TI FPD III Application Layer: YUV422 video output using the BT.656 digital video protocol 1024 X 768, 30 fps."
Camera Support: The system shall provide and manage the power supply to the camera. The system shall provide and manage the intrinsic calibration to the camera.
Vehicle CAN Interface: Physical Layer: Two-wire, termination (SAE J2284-3) Data Link Layer: 11898-1, ISO 11898-2 and ISO11898-6.

Performance
Test Configuration: Set the subject vehicle at the forward speed as 50 KM/h on the straight even road.
Test Case 1: Set the target vehicle at the forward speed as 50 + X1 KM/h to approach the subject vehicle from the left adjacent line, the expected output signal: CAN_BSD_Left_Alert (TTC) from the BSD module should be Y1 when the target vehicle's distance is Z1 meters from the subject vehicle's rear bumper.
Test Case 2: Set the target vehicle at the forward speed as 50 + X2 KM/h to approach the subject vehicle from the left cross line to the adjacent line when the target vehicle' head is at W1 meters from the subject vehicle's rear bumper, the expected output signal: CAN_BSD_Left_Alert (TTC) from the BSD module should be Y2 when the target vehicle's distance is Z2 meters from the subject vehicle's rear bumper.
Test Case 3: Set the target vehicle at the forward speed as 50 − X3 KM/h to approach the subject vehicle from the left adjacent line, the expected output signal: CAN_BSD_Left_Alert (TTC) from the BSD module should be Y3 when the target vehicle's distance is Z3 meters from the subject vehicle's rear bumper.

Table 3.3-3 BSD Functionality Requirement

Management Requirement:

The BSD system management requirements are listed here:

Management Requirement
The left camera BSD system development shall be ASPICE compliant.
The camera left Blind Spot Detection (BSD) system is a product of Automotive Safety integrity Level: ASIL B (D), and the development process shall be ISO 26262 compliant.
The BSD system shall be UN ECE 155 / 156 compliant.

Table 3.3-4 BSD Management Requirement

The BSD requirements above cover all aspects of an automotive ECU requirement engineering though they are not detailed enough to be implemented in a real project.

The requirements should be managed using the requirement management tools, such as the IBM DOORS, PTC Integrity. The suggested tool is the MS Excel like the one in Table 3.3-2 BSD Input & Output Signal, so that the input signals can be manipulated using the VBA script, which will make the system verification efficient.

The followings are the summary of Requirement Engineering:
- The main contents are for the feature function descriptions, which should be simple and straightforward, should be understood from the system description and the

input & output signal list.

- The diagnostic service functionalities are usually, as mentioned, outsourced to the dedicated third party, from the ECU developers' point of view, this part can be simplified as the request and response messages plus the required DID, RID and DTC that are project specific.

- The physical and electronic requirements should be either documented in the hardware requirements, or as simple requirement items in the system requirement document.

- About ISO 26262:
  - For the quality control aspects regarding to the development processes, the detailed requirements will be address by the organization development procedures, which should be verified during the safety assessment by following the Part 8 of ISO 26262.
  - The detailed performance measurement aspects regarding to the hardware Probabilistic Metrics, Single-Point Fault Metric, Latent Fault Metric, they will be address by the hardware development procedures and the hardware test cases.

- About the UN ECE 155 / 156 or other cybersecurity requirements: The cybersecurity requirements are the protections against external threat and attack, which do not trigger the system external behavior changes, so all the cybersecurity functionalities belong to the "internal functionalities", not the customer ones, however, the customers are the users of those functionalities. So, the efficient way to address the detailed requirements is to document them separately with the customer interfaces, which will be described in 3.4.2.5 Cybersecurity Function.

- Regarding to the AUTOSAR and other components that are from third parties, the integration verification criteria should be clarified about interfaces and the functionalities.

Check List 2 - Requirement Engineering

- Are the full system requirements including both technical and non-technical requirements defined?
- Can all the technical requirements be defined as the computer executable information (either the computer executable signal or the computer executable test cases)?
- For the computer executable signals, are all the relevant information (signal name, signal direction, signal length, value range, unit, accuracy, frequency) defined?
- For the computer executable test cases, are all the information (the signal to be tested, the test configuration, the test criteria) defined?
- Does the system description fully describe the system (interfaces, functionality and performance)?
- Are the system safety requirements defined?
- Are the system cybersecurity requirements defined?
- Are the system diagnostic requirements defined?
- Are the system AUTOSAR requirements defined?
- Are the system communication requirements defined?
- Are the system performance requirements defined?
- Are the system development management requirements defined?

3.4　System Architecture Design

Based on all the requirements that are collected and analyzed, the system under development can be described from the external description point of view or from the "black box" point of view as Figure 3.1-2 Computing System with Multiple Signals, and then system engineering needs to figure out how to realize such "black box", which is the content of System Architecture Design that will answer the following questions:

- How to derive the output data from the input data? (Which is the content of System Operation Concept)
- What functions are needed to derive those data? (Which is the content of System Structure Design)
- What devices are needed to derive those data? (Which is the content of System Electronic Architecture Design)
- What functions should be allocated on which devices and how they shall communicate with each other? (Which is the content of System Functionality Allocation)
- How does the output data timing is controlled? (Which are the contents of system latency design, system scheduling and the task arrangements)
- How to ensure those functions and devices' reliability? (Which is the content of FMEA)
- What safety measures are needed? (Which is the content of Safety)
- How to ensure those data are derived correctly and confidently and how to ensure those data are communicated with each other correctly and confidently? (Which is the content of Cybersecurity)
- How to make sure that the individual component work together correctly to establish the required product? (Which is the content of System Integration and Integration Verification)
- How to ensure that the system meets the environment and lifetime requirements? (Which is the content of System Black Box Verification)

3.4.1　System Operation Concept Design

The goal of system operation concept design is to figure out how to derive the required output signals from the given signals and the given conditions. And based on the information from the operation concept, the development will design the system structure including both hardware structure and software structure to ensure that the output signals meet the required timing that is the system latency design.

3.4.1.1　Derivation

To derive the output signals, there are commonly the following algorithms:
- Motion Control, such as Braking Control, Steering Control, Window Control, Seat Control, Doors Control.

The vehicle motion control can be derived using the PID algorithm.
- Information Control, such as Infotainment, Gateway Control, Wireless Control, Vehicle to X networking Control.

Those derivation logics can be derived based on the information communication protocol.

- Object Detection, such as Vehicle Front Camera, Rear Camera, BSD Camera, Front Radar, Rear Radar.

Those derivation logics can be derived from the object detection algorithm based on the camera pixel image, radar radio image.

The algorithms of derivations belong to the special feature development, which are product specific, not the contents of system engineering, nor is of this book. So, the derivation logics will be described using the generic formulas. Taking the generic system below as an example.

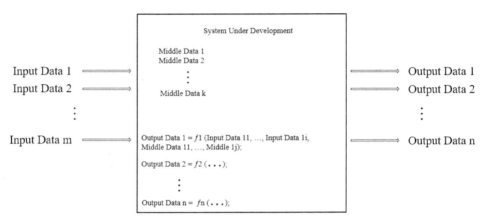

Figure 3.4-1 Computing System with Multiple Signals

Such system in Figure 3.4-1 Computing System with Multiple Signals above can be described by the formulas below:

Output Data 1 = $f1$ (Input Data 11, …, Input Data 1i, Middle Data 11, …, Middle Data 1j);

Output Data 2 = $f2$ (Input Data 21, …, Input Data 2l, Middle Data 21, …, Middle Data 2p);

…

Output Data n= fn (Input Data n1, …, Input Data nq, Middle Data 1n, …, Middle Data nr).

The method above can be done recursively to any data that need to be decomposed further into decompositions as the development progresses. For example, if the Middle Data 11 needs to be decomposed into such expression: Middle Data 11 = $fm11$ (Input Data 111, … Input Data 11i, Middle Data 111, …, Middle Data 11j), in which the $fm11$ is the calculation to derive the Middle data 11, the input data group of Input Data 111, …, Input Data 11i is a subset of input data group of Input Data 11, …, Input Data 1i, the middle data group of Middle Data 111, …, Middle Data 11j is a subset of middle data group of Middle Data 11, …, Middle Data 1j. Then the System Operation Concept method for the Middle Data 11 will be done by applying the System Operation Concept processes above to the expression of Middle Data 11 = $fm11$ (Input Data 111, … Input Data 11i, Middle Data 111, …, Middle Data 11j). The process of which can be illustrated in Figure 3.4-2 Decomposed Computing System.

3.4.1 System Operation Concept Design

The process is the system functionality decomposition, the purpose of which is to develop the information that is detailed enough for realization by either hardware or software. How detail the decomposition shall be will depend on the specific situation:

- Generally, the decomposition shall be such detail that the software coding can be done, or hardware structure can be implemented by following the information in the operation concept.
- If the realization will reuse the existed component, then the information regarding to the reuse shall be provided, such as the name, type, unit, resolution, range and default value.

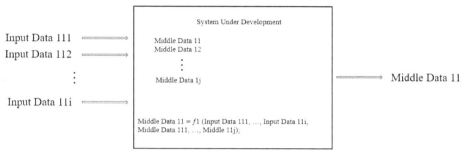

Figure 3.4-2 Decomposed Computing System

By establishing the operation concept above, the following information should be made available:

- The middle data definitions, including the data type, resolution and their characteristics.
- If the middle data are derived from other data, then the derivation functions and the needed data should be defined.
- If the data are received from other functions, partitions or devices, then the data receiving protocol should be defined including the hand-shaking protocols for data transfer, fault detection and fault recovery.
- The functional chain of events and behaviors should be defined, such as the function calls, interruptions.
- The logic sequence of data processing and control flow that are used to derive the results should be defined.

The information above is not only used to derive the results, but also used to design the system structure and the system reliability and safety.

An example of building the operation concept for a feature function will be described in 3.4.1.2 BSD Example, the following is another example of COM Send Signal which is used to integrate such software component.

The Com_SendSignal function is described in Table 3.4-1 COM Send Signal below that is one of functions in the COM module of AUTOSAR as an example to describe what the system operation concept looks like and how the decomposition process should be.

Service Name	Com_SendSignal

Interface Syntax	uint8 Com_SendSignal (
	    Com_SignalIdType SignalId,
	    const void* SignalDataPtr
	)

Sync / Async	Asynchronous
Reentrancy	Non-Reentrant for the same signal. Reentrant for different signals.
Parameters (in)	SignalId: Id of signal to be sent; SignalDataPtr: Reference to the signal data to be transmitted
Parameters (in & out)	None
Parameters (out)	None
Return value	uint8 E_OK: service has been accepted unit8 COM_SERVICE _NOT_AVAILABLE: corresponding I-PDU group was stopped (or service failed due to development error) uint8 COM_BUSY: in case the TP-Buffer is locked for large data types handling
Description	The service Com_SendSignal updates the signal object identified by Signal with the signal referenced by the SignalDataPtr parameter.

Table 3.4-1 COM Send Signal

If the system under development is to send a signal, then the required system can be described as:

Com_SendSignal_Status (Output_Stream) = Com_SendSignal (SignalId, SignalDataPtr);

In which, the Com_SendSignal_Status is the output data, Output_Stream is the output data stream that will depend on the communication protocol, such as CAN, FlexRay, Ethernet; the SignalId, SignalDataPtr are the input data, the Com_SendSignal is the relationship between the input data and the output data, which can be described as below to provide further detail information:

Com_SendSignal_Status (Output_Stream) = Com_SendSignal (SignalId, SignalDataPtr, COM_Protocol_ID, CRC, Data_Packet_Counter);

The formula above can be described as Figure 3.4-3 Com Send Signal, in which, furthermore, the middle data can be decomposed if needed, for example, the CRC can be decomposed to a function to calculate the CRC value, or can be an integrated function that calculates the CRC.

If the system under development is to integrate the component of Com_SendSignal from the AUTOSAR, then the integration information should be decomposed to such detail that it matches the contents in Table 3.4-1 COM Send Signal and the system description in Figure 3.4-3 Com Send Signal, as well.

3.4.1 System Operation Concept Design

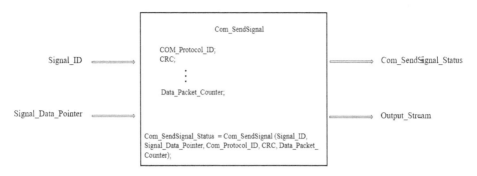

Figure 3.4-3 Com Send Signal

3.4.1.2 BSD Example

For the BSD example, based on the information in the section of 3.3 Requirement Engineering, the BSD system concept can be described as Figure 3.4-4 BSD Black Box View below:

Figure 3.4-4 BSD Black Box View

To facilitate the system operation concept design, the notation that is introduced in the section of 3.1.2 Notation is helpful describe the functional blocks, the execution sequence and the logic expressions, the design steps using the notation is below.

The whole BSD relationship can be expressed as the Executive Procedure Node (EPN) using the Data Driven system engineering (DDSE) notation in Figure 3.4-5 DDSE Notation of BDS Function below.

Among which, the signal: Time To Collision (TTC) is the required output signal, the signal: Object Detected is the object detection status, the signal of Camera Pixel Stream (Left_Video_LVDS) is the input signal, the signals: Object Relative Velocity, Object Relative Position, Object Tracking are the middle data. So, the output signal derivation can be described as:

Object_Detected (TTC) = f_detection_original (
    Left_Video_LVDS,
    Object_Relative_Position,
    Object_Relative_Velocity,
    Object_Tracking_State);

In the EPN mentioned, the system of Object Detection EPN is expressed using the comments sentence of C programming language that is assumed as the selected development programming language for the system under development.

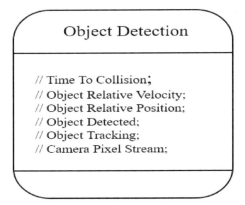

Figure 3.4-5 DDSE Notation of BDS Function

The EPN above can be further decomposed into sub-expressions or functionalities to have more detailed information of implementation.

In Figure 3.4-5 DDSE Notation of BDS Function above, the EPN of "Object Detection" is decomposed into the sub-EPNs: Camera Input Signal, Object Classification and Object Calculation in Figure 3.4-6 Decomposition of Object Detection below.

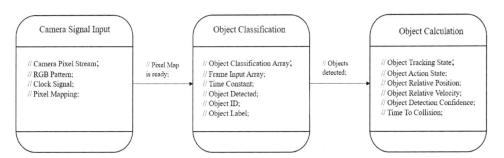

Figure 3.4-6 Decomposition of Object Detection

Among the decomposition, the Camera Input Signal EPN derives the camera pixel map as below:

Pixel_Mapping = f_mapping (Left_Video_LVDS, RGB_Pattern, Clock_Signal);

In which, the Left_Video_LVDS is the input camera pixel signals, both RGB_Pattern and Clock_Signal are design constants.

The Object Classification EPN derives the Object Detected, Object ID and Object Label as below:

Object_Detected (Object_ID, Object_Label) = f_classification (
    Pixel_Mapping,
    Object_Classification_Array,
    Frame_Input_Array,

Object_Detect_Time_Constant);
In which, the Pixel_Mapping is the input parameter from the previous EPN, the middle data Frame_Input_Array is derived from:
 Frame_Input_Array = f_frame (Pixel_Mapping, Frame_Input_Time_Constant);
both Object_Classification_Array and Object_Detect_Time_Constant are the design constants.

The Object Calculation EPN derives the Detection Confidence ratio, the Time To Collision (TTC) and the final output CAN signal: CAN_BSD_Left_Alert as below:
Object_Detection_Confidence = f_ratio (
        Object_ID,
        Object_Label,
        Object_Tracking_State,
        Object_Action_State);
In which, the Object_ID and Object_Label are the input parameters from previous EPN, Object_Action_State is the design constant, the middle data Object_Tracking_State is derived from:
Object_Tracking_State = f_tracking (Frame_Input_Array, Object_ID, Object_Label, Object_Traking_Time);

TTC = f_time (
        Object_Relative_Position,
        Object_Relative_Velocity,
        Object_Detection_Confidence);
In which, both Object_Relative_Position, Object_Relative_Velocity are middle data derived from below:
Object_Detected (Object_Relative_Position, Object_Relative_Velocity) = f_detection (
        Pixel_Mapping,
        Object_Classification_Array,
        Frame_Input_Array,
        Object_Detect_Time_Constant,
        CAN_Msg_Veh_Speed);

CAN_BSD_Left_Alert (TTC) = f_output (
        TTC,
        CAN_Msg_Ignition,
        CAN_Msg_Veh_Speed,
        ADC_Veh_Power,
        ADC_Env_Temperature);

In which, TTC is derived previously from another formula, and all other input parameters: CAN_Msg_Ignition, CAN_Msg_Veh_Speed, ADC_Veh_Power, ADC_Env_Temperature are the initial input signals.

So, by designing some middle results and middle data, the full BSD system operation concept can be derived as below:

3.4.1 System Operation Concept Design

Pixel_Mapping = *f*_mapping (
        Left_Video_LVDS,
        RGB_Pattern,
        Clock_Signal);
Frame_Input_Array = *f*_frame (
        Pixel_Mapping,
        Frame_Input_Time_Constant);
Object_Detected (Object_ID, Object_Label) = *f*_classification (
        Pixel_Mapping,
        Object_Classification_Array,
        Frame_Input_Array,
        Object_Detect_Time_Constant);
Object_Detected (Object_Relative_Position, Object_Relative_Velocity) = *f*_detection
(
        Pixel_Mapping,
        Object_Classification_Array,
        Frame_Input_Array,
        Object_Detect_Time_Constant,
        CAN_Msg_Veh_Speed);
Object_Detection_Confidence = *f*_ratio (
        Object_ID,
        Object_Label,
        Object_Tracking_State,
        Object_Action_State);
Object_Tracking_State = *f*_tracking (
        Frame_Input_Array,
        Object_ID,
        Object_Label,
        Object_Traking_Time);
TTC = *f*_time (
        Object_Relative_Position,
        Object_Relative_Velocity,
        Object_Detection_Confidence);
CAN_BSD_Left_Alert (TTC) = *f*_output (
        TTC,
        CAN_Msg_Ignition,
        CAN_Msg_Veh_Speed,
        ADC_Veh_Power,
        ADC_Env_Temperature);

In the relationships above, the following data are defined:
- Output signals:
  - TTC
  - CAN_BSD_Left_Alert (TTC)
- Input signals:
  - Left_Video_LVDS
  - CAN_Msg_Ignition

- o CAN_Msg_Veh_Speed
- o ADC_Veh_Power
- o ADC_Env_Temperature
- Middle Result:
  - o Pixel_Mapping
  - o RGB_Pattern
  - o Clock_Signal
  - o Object_Relative_Position
  - o Object_Relative_Velocity
  - o Object_Detected
  - o Object_Classification_Array
  - o Frame_Input_Array
  - o Frame_Input_Time_Constant
  - o Object_Detect_Time_Constant
  - o Object_ID
  - o Object_Label
  - o Object_Detection_Confidence
  - o Object_Tracking_State
  - o Object_Action_State
  - o Object_Traking_Time

All the information above is managed using the MS Excel sheet based on their relationships illustrated in Table 3.4-2 BSD System Operation Concept Design.

From the BSD operation concept above, the following information can be disclosed:
- All needed data and formulas to derive the required output signals

From the BSD operation concept, all the data and the formulas to derive the required output signal: CAN_BSD_Left_Alert (TTC) are disclosed, i.e., it needs the formula: f_output and its parameters, which, in turn, needs the formula: f_time and its parameters: Object_Detection_Confidence, Object_Relative_Position, Object_Relative_Velocity, …, in the end, the formula: Pixel_Mapping and its input signal: Left_Video_LVDS are needed.

- Operation and logic sequence

From the BSD operation concept, especially from the Table 3.4-2 BSD System Operation Concept Design, the data available sequence that is the middle data calculation sequence is disclosed. For example, before the calculation of Object_Detected (Object_Relative_Position, Object_Relative_Velocity) = $f\_detection$ can be executed, the parameters: Pixel_Mapping, Frame_Input_Array, CAN_Msg_Veh_Speed should be ready, and they need the Left_Video_LVDS receiving function and CAN transceiver function respectively.

Based on the operation and logic sequence, the functional chain of events and behaviors can be defined, such as the function calls, interruptions. In the BSD example above, before the $f\_detection$ is executed, all the needed data need to be ready, among them, the CAN_Msg_Veh_Speed is received from the vehicle CAN bus by the interruption routine, Pixel_Mapping and Frame_Input_Array are calculated from the input signals: camera pixel, so those data readiness needs to be synchronized that will be done by designing the task arrangement and scheduling.

Output Data	Middle Data	Middle Data	Middle Data	Middle Data	Input Data
CAN_BSD_Left_Alert (TTC)	TTC	Object_Relative_Position, Object_Relative_Velocity	Pixel_Mapping	Left_Video_LVDS	Left_Video_LVDS
			Pixel_Mapping	RGB_Pattern	
			Pixel_Mapping	Clock_Signal	
			Object_Classification_Array		
			Frame_Input_Array	Pixel_Mapping	
			Frame_Input_Array	Frame_Input_Time_Constant	
			Object_Detect_Time_Constant		
			CAN_Msg_Veh_Speed		
		Object_Detection_Confidence	Object_ID, Object_Label	Pixel_Mapping	
			Object_ID, Object_Label	Object_Classification_Array	
			Object_ID, Object_Label	Frame_Input_Array	
			Object_ID, Object_Label	Object_Detect_Time_Constant	
			Object_Action_State		
			Object_Tracking_State	Frame_Input_Array	
			Object_Tracking_State	Object_ID, Object_Label	
			Object_Tracking_State	Object_Traking_Time	
	CAN_Msg_Ignition				
	CAN_Msg_Veh_Speed				
	ADC_Veh_Power				
	ADC_ Env_Temperature				

Table 3.4-2 BSD System Operation Concept Design

- Operation and data characteristics

From the BSD operation concept, the operation and data type, resolution and their characteristics are disclosed. For example, from the formula below:

Object_Detected (Object_Relative_Position, Object_Relative_Velocity) = f_detection
(

   Pixel_Mapping,
Object_Classification_Array,
Frame_Input_Array,
Object_Detect_Time_Constant,
CAN_Msg_Veh_Speed);

the f_detection defines the data type, resolution and the characteristics for all the data in it, such as the Pixel_Mapping is a matrix data derived from the input signal Left_Video_LVDS, which represents the environment sensed by the camera, the Object_Classification_Array is a matrix data as well that is used as the pattern to figure out the objects in the camera image data. The characteristics such as the matrix data need the processor to have the special capacity to handle such data, which will help the microcontroller selection in the hardware design.

- The needed devices and protocols

In the BSD example above, the input data: CAN_Msg_Veh_Speed and CAN_Msg_Ignition are received from, and the output signal: CAN_BSD_Left_Alert (TTC) is transmitted to the vehicle CAN bus, and to do which, the device that is the protocol CAN compliant including both hardware and software is needed for all information about the CAN signal communication.

And for the calculation: Pixel_Mapping = f_mapping (Left_Video_LVDS, RGB_Pattern, Clock_Signal) which needs the device to receive the camera input pixel signals: Left_Video_LVDS that is the Data Link Layer protocol: TI FPD III compliant.

All the information defined by the system operation concept will be used for the next development: design the system structure, system hardware architecture, safety development and system verification, and all the next developments are to realize the system operation concept.

3.4.1 System Operation Concept Design

Check List 3 - System Operation Concept Design

- Are all the system output data derived from the system operation concept?
- Are all the system input data used in the system operation concept?
- Are all the needed calculations, data (data types, data characteristics, data resolutions, data flows, relationships) and middle data derivations defined?
- Are the definitions above detailed enough to be allocated to the specific devices or components (for example, some data calculations need the matrix operation ability or the image process ability)?
- Are all the process sequences and logics defined?
- Are the derivations in the system operation concept simulated to prove that they are correct and feasible?

3.4.2 System Structure Design

The System Operation Concept Design provides the solution of how to derive the output data from the input data via certain design middle data, then the next step of development is to structure the platform to realize the functionalities that are needed in the system operation concept, which will focus on what functions and what devices are needed to derive those data and how they should cooperate with each other.

The targets of system structure design are:
- The calculations in the operation concept need to be realized.
- The input data need to be imported from the input devices, and the middle data need to be imported from some other component functions.
- Some of data need to be stored in the external NVM to be used in the next power cycle.
- All data need to be stored in the RAM during runtime, which needs to be managed.
- The used devices including the processor need to be managed, such as which tasks need to be run first and how long they need to run.
- All those management services above need to communicate with each other.

To realize the operation concept, the functions and the devices to provide the services are needed, and the extra functions and services are needed to manage those functions and devices, all of the functions, devices and services that support the system operation concept need to be designed and structured into the system infrastructure.

A lot of effort of development is needed to design and implement an infrastructure in a computing system, fortunately, in the automotive industry, most of the services and managements are done by the AUTOSAR.

For the reasons of product quality and development efficiency, almost all automotive ECUs are required to be designed based on the AUTOSAR structure, except the ones that have long development history and complicate functions, such as the Electronic Brake Control Module (EBCM), Engie Control Module (ECM).

The benefits of using the AUTOSAR are that all the infrastructure functions can be carried over from the AUTOSAR:
- System Management Service, such as memory partition management, communication between those partitions, communication between the system service components and the peripherals, time or scheduling management, shut down and start up.
- Memory device management, such as the external RAM, external Flash.
- Crypto Service management, such as the symmetric encryption and decryption (AES) and asymmetry encryption and decryption (RSA), Secure Hash Algorithm (SHA).
- Communication management, such as CAN, Ethernet, FlexRay, TCP/IP.
- Networking Management, such Wireless, Partial Network.
- Special Device Integration using the Complex Device Driver (CDD), such as Camera, Radar.

All the automotive ECU designed based on the AUTOSAR has the basic structure illustrated in Figure 3.4-7 ECU Component Structure below. Based on such structure, the system under development focuses on only the Application Layer that commonly consists of the following components:

- Feature Function
- Application Mode Manager
  o Application State Management
  o Reliability Monitor
- Serial Signal Manager
- Diagnostic Services
- Cybersecurity Function

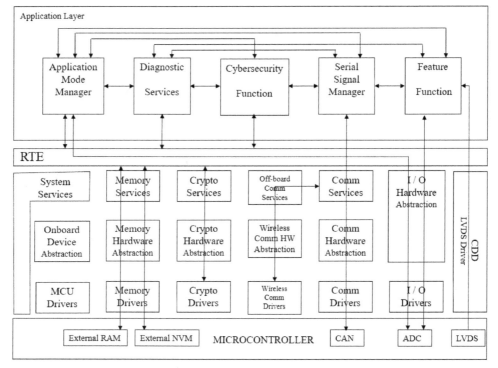

Figure 3.4-7 ECU Component Structure

In which, the Feature Function is the required ECU specific functions, and the other components: Application Mode Manager, Serial Signal Manager, Diagnostic Services and the Cybersecurity Function are the bridge Functions between the Feature Function and the AUTOSAR components, their functionalities are common for all the automotive ECUs though the implementations and configurations are project specific, i.e. all ECUs have those functions, but the realizations are somehow different for the individual ECU depending on the project requirements.

In AUTOSAR architecture, there are only two types of communication interfaces, which are used in this book, as well:

- Sender / Receiver Interface: The sender sends the messages to the receiver.
- Client / Server Interface: The client invokes the functions in the server.

In the AUTOSAR terminology, all the components in the application layer are called Software Components (SW-C), and the RTE in Figure 3.4-7 ECU Component Structure above is an AUTOSAR component that is most close to the SW-Cs, which is the Functional Virtual Bus (VFB) to facilitate the communication between the components in the AUTOSAR and the application layer. Based on which, the typical data input and output illustrated in Figure 2.2-2 ECU Input and Output Signals can be described in Figure 3.4-7 ECU Component Structure, The interfaces and contents of which will be described in the next sections.

3.4.2.1 Feature Function

The feature functions represent the required characteristics of the specific ECUs, such as:
- How the radar ECU detects the object(s) from the radio signals that are transmitted from the ECU antennas and reflected from the environment back to the ECU.
- How the camera ECU detects the object(s) from the environment that it can "see" through its lens.
- How the brake ECU controls the braking motor to output the braking force based on the brake pedal, vehicle velocity and trajectory.

The approach to the feature function is the system operation concept design that is introduced in the section of 3.4.1 System Operation Concept Design, which should fully reveal the derivations of feature function output signals including the used data type, resolution, characteristics and the relationships between the data.

The feature functions usually need to communicate with the Complex Device Driver (CDD) via the RTE of AUTOSAR to have the special signals for the feature functions, such as the camera ECU needs the camera pixel stream signals, the radar ECU needs the radio signals, the brake ECU needs the brake pedal signal, vehicle trajectory signal. Those feature functions and their input and output signals must be handled by the ECU supplier because they are project specific.

Some other feature functions, such as Door Control Module, Power Liftgate Control Module, need the microcontroller built-in peripherals in the microcontroller to input and output the signals, such as the PWM for motion control, digital signal for position, which can be handled using the device driver that is from the microcontroller supplier.

Those devices used by the feature functions need the special developments about:
- Calibration: to achieve the designed feature functionalities, those devices are needed to be calibrated every time when the ECU is power on and sometimes during the ECU's normal operation.
- Configuration: to be reused for variant projects, the devices are needed to be configured for different circumstances.
- Self-test: To work reliably, the devices must be self-tested during the power on and sometimes during the normal operation, or on demand.
- Shutdown handling: most of actuators need to be positioned at the designed location at the time of shutdown.

Because of the activities above, those devices need more time during the power on initialization phase and the shutdown phase, so, the application mode manager that is responsible for those phases needs to consider such features, and it is better to use asynchronous function calls to do the initializations and shutdown requests to the feature functions.

Taking the BSD system in the section of 3.1.3 Example as an example, whose input signal is the camera signal input. In which, the environment sensing sensor is a camera which captures the environment using pictures that then are transferred to a stream pixel.

To transfer the stream pixel from the location where the camera is installed to the location where the BSD module is installed, the serializer in the camera and the de-serializer in the BSD module should be used. Then the stream pixel received by the de-serializer is transferred to the microcontroller, in which the CDD in turn transfers the stream pixel to the feature functions that usually are allocated in one of Application Processing Unit (APU) cores that will be used to process the pixels to figure out if there is an object detected.

The feature function input and output data flows are described in Figure 3.4-8 Feature Function I/O Flow below.

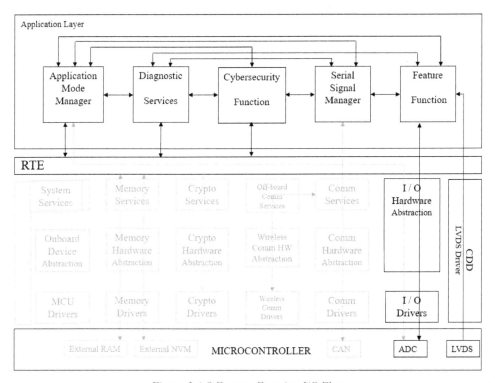

Figure 3.4-8 Feature Function I/O Flow

In addition, the feature functions need to collaborate with other application layer components illustrated in Figure 3.4-8 Feature Function I/O Flow above to operate on the platform, among which the main interactions are in Table 3.4-3 Feature Function Interface below:

Feature Function Interface (Legend: App-> Application Mode Manager, Diag->Diagnostic Service, Seri->Serial Signal Manager, Cyber->Cybersecurity, Feat->Feature Function, AUTO-> AUTOSAR)				
Interface Signal	**Source**	**Target**	**Type**	**Content**
Execution	App	Feat	Sender / Receiver	The application Mode Manager sends out the execution command to the feature function
Shutdown Request	App	Feat	Sender / Receiver	The application Mode Manager sends out the shutdown request to all the feature function
Filtered Voltage	App	Feat	Sender / Receiver	The application Mode Manager sends out the filtered supply voltage value to the feature function
Reset Request	Feat	App	Sender / Receiver	The feature function sends out the reset request to application Mode Manager
Diagnostic Event Notification	AUTO	Feat	Sender / Receiver	The AUTOSAR sends the diagnostic event notification to the feature function, such as the DTC inhibit, DTC recovery.
Error Handling Request	Feat	AUTO	Client / Server	The feature function sends out the error information to AUTO
Diagnostic Service Request	Diag	Feat	Client / Server	The diagnostic service component sends the diagnostic service request to feature function component, such as the internal information query, internal information query.
Serial Message	Seri	Feat	Sender / Receiver	The serial signal manager sends the serial signal to the feature function
Event Information	Feat	Seri	Sender / Receiver	The feature function send the event information to serial signal manager for the serial communication

Table 3.4-3 Feature Function Interface

- Application Mode Manager: The application mode manager controls the startup and shutdown of the feature functions, and provides the environment information about the vehicle power supply voltage which is critical to some feature functions that are related to the motor control.
- Serial Signal Manager: It processes the serial communication input and output signals by checking the integrity and validity, if needed, and ensures the

communication safety and security.
- Diagnostic Service component provides the maintenance functions, such as information query, parameter update and maintenance routine executions.

To collaborate with other components in the vehicle, every automotive ECU feature function must have two kinds of enabling conditions: Vehicle Status and Environment Status.

Vehicle Status: Which are the vehicle conditions that enable the required feature functions. Those conditions are the feature functions' enablers, i.e., if those conditions are satisfied, then the feature functions are enabled, otherwise, the feature functions will be disabled.

In the automotive ECUs, almost all of feature functions have the following two vehicle enabling conditions:
- Vehicle Ignition Status
- Vehicle Speed

Those two conditions are to make sure that the vehicle is in the moving status. For some other feature functions, the vehicle transmission status is another key enabling condition, such as the Rear-View Camera, Backing Automatic Braking, etc., those functions will work only if the transmissions gear is in the Reverse status. But it is not for the BSD.

The enabling signals are normally about the subject vehicle status, so they come from the vehicle communication bus, such CAN bus or ethernet bus. In the BSD example, those two enabling signal processes are expressed in the EPN: Ignition Enabler, EPN: Speed Enabler respectively in Figure 3.4-9 BSD Enabler below.

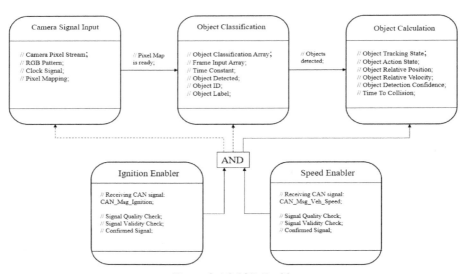

Figure 3.4-9 BSD Enabler

It depends on the feature functions' characters to decide where for the enablers to function, i.e.
- If the enablers' conditions are not satisfied, and the feature functions should be stopped at the beginning where the input signals are processed, and neither current input signals nor middle results are useful for the future process, then the enablers should be checked at the beginning. This case is illustrated using the dash line from

the enabler to the EPN: Camera Signal Input.

- If the enablers' conditions are not satisfied, however, either the input signals or the middle results are still useful for the future process, then the enablers should be checked at the middle. This case is illustrated using the dash line from the enabler to the EPN: Object Classification.
- If the enablers' conditions are not satisfied, however, all the input signals and the middle results are still useful for the future process, i.e., the output signals are dependable on the signals' processing history, then the enablers should be checked at the end where the output signals are output. This case is illustrated using the solid line from the enabler to the EPN: Object Calculation.

For the BSD example, it belongs to the last case above, i.e., the required output signal: CAN_BSD_Left_Alert (TTC) depends on the input signals and middle results' history, such as the Object Tracking State that is tracking the object history in the detection zone, so all the input signals and the middle result are still useful even if the BSD output signals are disabled. So, the enablers control the Object Detection EPN only that is illustrated by the solid line.

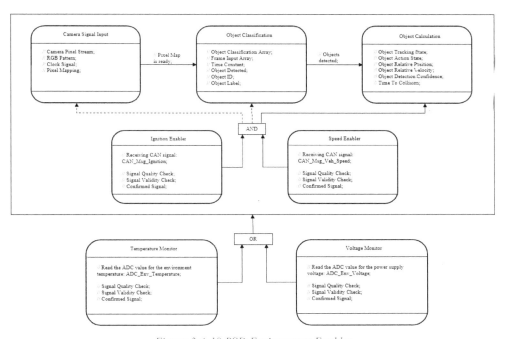

Figure 3.4-10 BSD Environment Enabler

Environment Status: In an automotive ECU, another enabler is the Environment Enabler, all ECUs need to monitor the environment where they are working in about two factors:

- The Environment Temperature: the ECUs should work only in certain temperature range, such as from -40°C to +80°C, if the temperature is out of the range, then the electronic components in the ECUs may not work reliably.
- The Power Supply Voltage from the Vehicle: the ECUs should work only in certain voltage range, such as from 8V to 16V, if the power supply voltage from the vehicle is out of the range, then the electronic components in the ECUs may not

work reliably.

Both factors above are measured using the Analog to Digital Converter (ADC) that is built in the microcontroller. Since they are impact the whole ECU, so the control logic is illustrated in Figure 3.4-10 BSD Environment Enabler.

Note: the Environment Enabler usually is part of the Application Mode Manger in the section below, for the integrity reason, it is introduced in this section.

3.4.2.2   Application Mode Manager

Application Mode Manager is the component that manages all the components' operation modes in the application layer to keep the ECU in the good "health" state.

In a computing system including an automotive ECU, both hardware components and software components have various operational modes based on the power supply status and their internal health status.

From hardware point of view, the hardware components, illustrated in Figure 2.2-2 ECU Input and Output Signals are the power supply, the COM interfaces, the external RAM and NVM and the microcontroller including the RPU, APU, GPU, FPGA, SHE and internal EEPROM and RAM, have the unreliable status during the power on and power off because those electronic elements need to be energized or de-energized during the time, so their functionalities may not be trustable. And after the power on, the devices need to be configured to function as the design, and some of them need to be calibrated, which will take some time for those configurations to be effective and for the calibration to be finished.

From software point of view, after the microcontroller power on, the software needs to be loaded from the either internal or external NVM into RAM to be executable by the processor, after which, the data need to be initialized, during which, the software will not be able to operate as normal.

And even during the normal operation, the system needs to monitor both hardware and software health status to make sure the product to operate correctly. If there is something critically wrong that will impact the system reliability, the system needs to stop the normal operation and go to fault states, in which the system cannot operate normally.

The situations above need to be handled for every hardware and software component, so it is common that there is the dedicated function for such handlings. Each component has its own structure, configurations, so the handling activities are component specific. The individual component needs to collaborate with each other to perform the system functionalities, to do so, in the application layer the tasks to collaborate the components and handle those situations are allocated to a functional component called: Application Mode Manager.

The Application Mode Manager implements the following functionalities:
- ECU power on and power off
- Platform Hardware health status monitoring and aggregation, such as I/O ports, RAM, EEPROM, NVM, etc.
- Platform Software health status monitoring and aggregation, such as, software partitions' start and stop, reprogramming request, execution sequence monitoring using the watchdogs.

- ECU environment monitoring and aggregation, such as temperature and vehicle power supply voltage monitoring.
- ECU cybersecurity Intrusion Detection monitoring
- ECU reset monitoring

To do the tasks above, the Application Mode Manager collaborates with both following components in the AUTOSAR:

- the ECU State Manager (EcuM), which manages the ECU operation mode.
- the Basic Software Mode Manager (BswM), which manages the BSWs operation mode.

The main responsibilities of Application Mode Manager are for the startup process and shutdown process, the process targets could be:

- Some application layer component(s)
- Some device(s)
- Partition(s)
- Processor(s)
- The ECU

And for the integrity and conveniency reasons, the monitoring activities that may result in the shutdown process are implemented in it.

During the startup process, the EcuM validates the wakeup signal when a wakeup from sleep or battery connect occurs, then initializes a configurable set of BSWs, and then hands over control to BswM once initialization is complete. BswM is configured to finish initialization by starting the Schedule Manager and RTE, and any other configured BSWs that are not initialized by EcuM. At this point, the ECU can execute its application SW-Cs.

During the shutdown process, the BswM requests the mode switch to the EcuM to perform the shutdown after the BSWs finish the shutdown preparation. However, BswM does not have any AUTOSAR defined function to determine when the ECU needs to transit to the shutdown and what shutdown target state (Sleep, Off, or Reset) should be reached, which should be done by the Application Mode Manger.

An automotive ECU does not need to communicate with all other ECUs in the vehicle network illustrated in Figure 2.2-1 Vehicle ECU Network, rather it needs only to communicate with a few of them to do certain functionalities. The ECUs related to certain functionalities in the vehicle network form the Network Management Clusters or Partial Network, i.e., an ECU needs only to communicate with the ECUs in its Network Management Cluster to perform its designed functionality, so that, if there is a function that should run, then only the relative ECUs should stay in power on, others that are not relevant should go to the bus-sleep mode to save the battery power.

The vehicle network management activities in an ECU will be done by the Application Mode Manager by cooperating with the Serial Signal Manager.

An automotive ECU's power state depends on its role in the vehicle network. The role can be either active role ECU or passive role ECU:

- The active role ECU needs other ECUs in the vehicle network together to perform the designed functionalities. Such ECUs can wake up the vehicle network. For example, for the functionality of remotely activating the power liftgate when the vehicle is ignition off, which is initiated by pressing the liftgate bottom on the keyfob, then the wireless ECU will inform the vehicle body controller module (BCM) that in turn send the liftgate command to the power liftgate module (PLG) to activate the gate. In this functionality, the wireless ECU is an active role ECU. The active role ECU can be wakened up locally.

- The passive role ECU cannot wake up the vehicle network, it can only be wakened up by others to perform certain functionalities. In the example above, the power liftgate module (PLG) is a passive role ECU. And both the camera ECU and radar ECU in the section of 3.1.3 Example are the passive role ECUs that in the BSD functionality cluster

Such vehicle network management cluster activation and deactivation is managed by the Network Management (NM) component in the AUTOSAR, which needs to be configured for the specific project to coordinate the network sleep – wake up strategy.

The AUTOSAR NM coordination algorithm is based on periodic NM messages, which are received by all ECUs in the cluster via broadcast transmission. Reception of NM messages indicates that sending ECUs want to keep the NM-cluster awake. If any ECU is ready to go to the Bus-Sleep Mode, it stops sending NM messages, but as long as NM messages from other nodes are received, it postpones transition to the Bus-Sleep Mode. Finally, if a configured timer elapses because no NM message are received anymore, every ECUs in the NM-cluster initiates transition to the Bus-Sleep Mode, which is illustrated in Figure 3.4-11 Network Mode Diagram below.

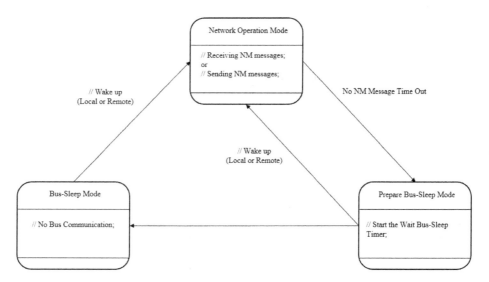

Figure 3.4-11 Network Mode Diagram

If any ECU in the NM-cluster requires others cooperation using the communication bus, it can wake-up the NM-cluster from the Bus-Sleep Mode by transmitting NM messages:

- Every network ECU shall transmit periodic NM messages as long as it requires bus-communication; otherwise, it shall not transmit any NM messages.
- If none of ECU uses the communication bus, and there is not any NM message on

the bus for a configurable amount of time, then the network should go to bus-sleep mode.

The AUTOSAR NM component communicates the network status to the BswM, which in turn cooperates with the application mode manager to manage the application layer components.

The Application Mode Manager is a standardized implementation to corporate with BswM to perform the application layer operation management, from the point of which, the Application Mode Manger is the extension of BswM, however, all the functions in the application layer including the Application Mode Manger are project specific, so, to increase the modulization and re-use ability, the Application Mode Manager is implemented as a SW-C.

To collaborate with the BswM and EcuM, the Application Mode Manager needs to implement the states in Figure 3.4-12 State Flow in Mode Manager below, in which the following activities are executed.

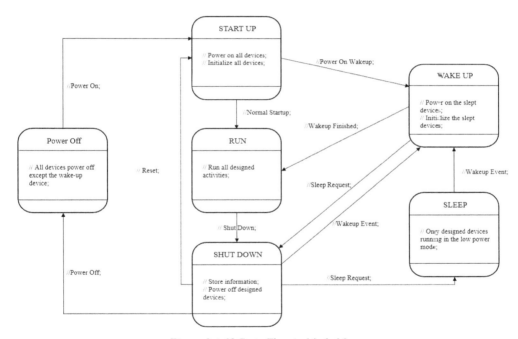

Figure 3.4-12 State Flow in Mode Manager

START UP and WAKE UP States: During the startup and wake up processes, the Application Mode Manager needs to do the following:

- Disable the vehicle network transmission communication.
- Power on and initialize the project specific hardware devices that are needed by the feature functions, such as the camera input pixel stream device (deserializer), LVDS, the radar radio signal device (antenna), power supply unit, ADC, etc. The other common device, such as RAM, ROM, Flash, EEPROM, CAN or other serial communication interfaces, can be initialized by the drivers in the AUTOSAR.

- Initialize all components in the Application Layer: Application Mode Manger (itself), Feature Function, Serial Signal Manager, Diagnostic Services, Cybersecurity Function by setting the initial values, reading the contents from the NVM, calibrating and configuring the application layer functions.
- After that, if all initialization checking is passed and if the BswM is in RUN state, then the Application Mode Manage should give the permission for all the application layer components to run.

SLEEP and SHUTDOWN States: The main function of the Application Mode Manager is to determine when there are no longer requests to run any functions in the application layer components. When this situation occurs, the ECU should go to a low power state to save power, either off or sleep state which is project specific depending on HW design. To do so, the application mode manager sends out the mode switch requests to the BswM to change the ECU operation state.

During the shutdown process, the application mode manager needs to:
- Disable the vehicle network transmission communication.
- Monitor each application layer components' status to determine if all of them are ready to go to shutdown state, because all or some of them need to do some preparations, such as saving the current status data and middle results data into the NVM to be used in the next function cycle, moving the actuators to the proper position.
- Monitor if there are the wakeup signals occur that should cause the ECU back to normal operation state from the shutdown process.
- Monitor if the shutdown conditions are satisfied, usually one of the ECUs' sleep conditions is that the vehicle network where the ECU is located is in silence for certain time.
- Request the BswM to initiate a shutdown.

Platform Status Monitor in all states: Another main function of the Application Mode Manger is to monitor the platform error and health status, and inform all the SW-Cs about the status, in most of cases, take the appropriate recovery actions if the issues impact the functions.

There are two types issues to be monitored:
- Unhealthy behavior, which is something suspicious, such as, too many reset requests from the external device.
- Error, which is something wrong

Both of issues above are called "Fault".

Faults can occur in both development process that are the development mistakes, and operation that are the systematic faults, and the handling of systematic faults consist of:
- Fault Detection
- Fault Isolation
- Fault Recovery

The mechanism of such fault handling distributes in the whole ECU system, which will be described in the sections of FMEA and Safety in detail. The relationship between the fault handling mechanism and the application mode manager is that the mechanism detects the faults and react to the faults, and the application mode manager will decide the application layer operation mode based on the faults. The application mode manager covers below aspects:

- Platform Hardware health status monitoring

To achieve the reliable operation, all devices in the ECU need be checked periodically to ensure that they are working properly. Some of the checks can be done in the AUTODAR by the devices' drivers, some of them, especially the project specific devices, such as camera signal input device (serializer and deserializer), radar antenna, ADC and DAC, need to be handled by the SW-Cs.

In some cases, especially to achieve certain safety requirements, the extra checks are needed for some devices, such as to satisfy the ASIL B in the QM devices, such as APU cores, external RAM, ROM, EEPROM, Flash, the periodically checking about the contents' integrity is necessary.

Once a hardware issue is detected, the Application Mode Manager needs to recover the platform from the issue, which usually is done by the resetting the ECU.

- Platform Software health status monitoring

The type of monitoring includes:
  - software aliveness monitoring or software execution monitoring using the watchdogs
  - BSW state and errors monitoring
  - communication session monitoring, such as: DTC clear operations, reprogramming requests, seed/key access failures that exceed design thresholds in a designed time window.

- ECU environment monitoring

All ECUs need to monitor the environment where they are working in about two factors that are the conditions for the environment enabler introduced in the feature function above:
  - The Environment Temperature: the ECUs should work only in certain temperature range, such as from -40°C to +80°C, if the temperature is out of the range, then the electronic components in the ECUs may not work reliably.
  - The Power Supply Voltage from the Vehicle: the ECUs should work only in certain voltage range, such as from 8V to 16V, if the power supply voltage from the vehicle is out of the range, then the electronic components in the ECUs may not work reliably.
    Both factors above are measured using the Analog to Digital Converter (ADC) that is built in the microcontroller.

The data flows that the Application Mode manager interacts with other application layer components, and it inputs the environment information data flow are described in Figure 3.4-13 Application Mode Manager Data Flow below.

3.4.2 System Structure Design

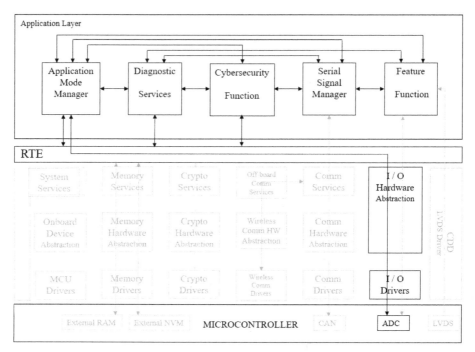

Figure 3.4-13 Application Mode Manager Data Flow

Based on the information above, the suggested interfaces for the Application Mode Manager software component are below:

Application Mode Manager Interface				
(Legend: App-> Application Mode Manager, Diag->Diagnostic Service, Seri->Serial Signal Manager, Cyber->Cybersecurity, Feat->Feature Function, AUTO-> AUTOSAR)				
Interface Signal	**Source**	**Target**	**Type**	**Content**
Execution	App	Diag Cyber Seri Feat	Sender / Receiver	The application Mode Manager sends out the execution command to all the application layer components
Shutdown Request	App	Diag Cyber Seri Feat	Sender / Receiver	The application Mode Manager sends out the shutdown request to all the application layer components
Operation Mode Request	App	AUTO	Sender / Receiver	The application Mode Manager sends out the operation mode request to AUTOSAR to change the operation mode
Error Handling Request	App	AUTO	Client / Server	The application Mode Manager sends out the error information to AUTO

Filtered Voltage	App	Diag Cyber Seri Feat AUTO	Sender / Receiver	The application Mode Manager sends out the filtered supply voltage value to all the application layer components and the AUTOSAR
Event Information	App	Seri	Sender / Receiver	The application mode manager sends the event information to serial signal manager to communicate with other ECUs, such as the subject ECU status.
Reset Request	Diag Cyber Seri Feat AUTO	App	Sender / Receiver	The application layer components and the AUTOSAR sends out the reset request to application Mode Manager
Current Operation Mode	AUTO	App	Sender / Receiver	The AUTOSAR sends the current operation mode to inform the application mode manager about the current operation mode
Measured Voltage	AUTO	App	Sender / Receiver	The ADC in AUTOSAR sends the measured voltage value to the application mode manager
Measured Temperature	AUTO	App	Sender / Receiver	The ADC in AUTOSAR sends the measured temperature value to the application mode manager
Diagnostic Event Notification	AUTO	App	Sender / Receiver	The AUTOSAR sends the diagnostic event notification about the error handling to the application mode manager to enable the application layer enter the safe state, and bout the AUTOSAR internal status information.
Diagnostic Service Request	Diag	App	Client / Server	The Diagnostic Service component sends the diagnostic service request to the application mode manager, such as current status, DTC setting inhibit, communication inhibit, internal parameter values.
Serial Message	Seri	App	Sender / Receiver	The serial signal manager sends the serial signal to the application mode manager

Table 3.4-4 Application Mode Manager Interface

3.4.2.3 Serial Signal Manager

The Serial Signal Manager is to process the serial communication signals between the subject ECU and the host vehicle to do two tasks:

- Communicate with each other about the working status, all of which is from the vehicle serial communication bus like ethernet, CAN, Ethernet, FlexRay.
- Receive and respond to the diagnostic requests, which will be described in detail in the section of 3.4.2.4 Diagnostic Service.

The serial signal manager's data flow and the interactions with other application layer components are illustrated in Figure 3.4-14 Serial Signal Manager Data Flow below.

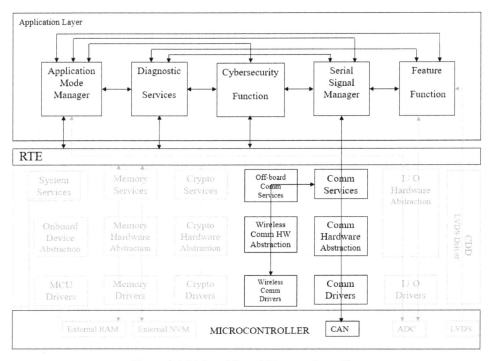

Figure 3.4-14 Serial Signal Manager Data Flow

By knowing the vehicle status, the ECU can decide if the feature functions should be engaged. So, such vehicle status signals are the Feature Functionality Enabler that are used in the section of 3.4.2.1 Feature Function, on the other hand, the ECU needs to inform the host vehicle about its working status, as well. The working status signals include both overview status and quantitative status signals, for example in the BSD system:

- BSD receives the vehicle dynamic overview status signal that is the ignition status.
- BSD receives the vehicle quantitative dynamic status signal that is the vehicle speed.
- BSD sends out its own overview object detection status that is the Object_Detected.
- BSD sends out its own quantitative object detection status that is the TTC.
- BSD sends out its own fault status if it is in a fault status, such as the environment temperature is out of working range.

The working status signals are safety and cybersecurity critical if applicable.

All the serial communications used between the ECUs in an automotive are asynchronous serial communication that do not have a clocking signal, instead they rely on unique Start and Stop bits to separate data packets, which is in contrast to synchronous serial communications that are used in Serial Peripheral Interface (SPI) and the Inter-Integrated Circuit (I2C) protocol, which have a clocking signal to synchronize the data transmission, for example, the memory chips like Static Random-Access Memory (SRAM), Electrically Erasable Programmable Read-only Memory (EEPROM) are common components that use synchronous communication.

The serial signal manager needs to manage its own working state by cooperating with the application mode manager, which consists:

- Initial State

Once the serial signal manager is started to execute, it should initialize its internal variables, configure the parameters, reload the stored information from the NVM, especially initialize the receiving and the transmitting contents by using the default values. After the initialization is done, it needs to inform the other application layer components about its ready to work status.

- Shutdown State

Before the shutdown, the serial manager needs to store the internal information, especially the status information that will be used in the next execution cycle into the NVM, after which is done, it needs to inform the application mode manager that it is ready to shut down.

- Disabled State

If some critical faults occur in the system that will hazard the system's functionalities, especially the safety related functionalities, then one of the system reactions to the faults is transit into the designed safe state, in which the system either disable the serial communication or set the serial signal value as Signal Not Available (SNA) or Invalid to warn the host vehicle, which should be as soon as possible to short the fault reaction time.

- Normal State

In the normal state, the basic tasks of serial signal manager are to execute the activities of transmitting and receiving signals, for which the major task is checking the integrity and validity.

Integrity:

For the received signals, the serial signal manager should check if the signal values are not changed during the transmission caused by the transmission errors, which can be done at two levels: the device drive level and application level.

At the device driver level, all the serial communication protocols have the cyclic redundancy check (CRC) value attached to the signals to check the data integrity, which is done by the serial communication device either hardware or software driver, which is handled by the AUTOSAR. The serial signal manager needs to cooperate with AUTOSAR to handle the checking status.

At the application level, the sender application software can calculate the signals' checksum, then integrate the values into the signals, then the receiver software needs to recalculate the checksum based on received signals to check if the receiver's checksum matches the sender's one.

For the transmitted signals, the serial signal manager should build the integrity check values into the signals, such as check sum or other protection values, such as some encryption values.

Validity:
Validity checking further is divided into the following check:

Availability Check: if the senders are not in normal operation status, such as in the initialization state, in fault state, then the sender will send the value of Signal Not Available (SNA) or Invalid to indicate the signal's availability, which is the "handshaking" of signal contents that should be carefully designed by the vehicle system designers.

Timing Check: the signal timing check can be done only for the periodic signals, for which, the receiver needs to check if the periodic signal is received following the required time range defined by the communication protocol. If not, the received value should be discarded. The timing issue can be caused either by the transmission path or the signals' sender, and the issues are classified as either Signal Time Out or Lost Communication:

- Signal Time Out is a temporary issue which can be detected by checking the time received against the designed periodic time, to handle which, the receiver can discard the received signal then use either the default value or previously received valid value, the strategy will depend on the design.
  Signal Time Out detection can make use the AUTOSAR Callbacks from the communication stack callback function of deadline monitoring ComTimeoutNotification: Com_CbkRxTOut() and message received notification: Com_CbkRxAck().

- Lost communication is a permanent issue which can be detected by the detection of continuous Signal Time Out issues, to handle which, the receiver must report the issue to the error handling software component that is generally the DEM in AUTOSAR, then usually use the value of Signal Not Available (SNA) to indicate such situation to all the signal consumer software components in the application layer, the strategy will depend on the design.

Authentication Check: the authentication check is applicable to the cybersecurity application, which is to check if the signal is authenticated by the dedicated sender consisting of:

- Message Rolling Counter (MRC) is a protection to against the message rolling back attack that the attacker will attack the system by sending the valid messages repeatedly, which will mislead the system behaviors or impact the system's communication capacity. The MRC is to provide a mechanism to increment the MRC based on the messages' transmission, and then the MRC will be integrated into the messages, and the receivers need to maintain receiving MRC to check against the counter values from the senders to detect the rolling back attack. The Message Rolling Counter mechanism needs to have the synchronization between the sender and receiver's MRC counters in the cases of first-time communication, recovery from communication stop by faults.

- Message Encryption: The confidential information transmitted in the serial communication can be encrypted by the sender, and then decrypted by the receiver to protect the information from the unintended disclose and modification, the detail of which will be described in the section of 3.4.7 Cybersecurity.

The Serial Signal Manager component uses the following interfaces to interact with other components:

Serial Signal Manager Interface
(Legend: App-> Application Mode Manager, Diag->Diagnostic Service, Seri->Serial Signal Manager, Cyber->Cybersecurity, Feat->Feature Function, AUTO->AUTOSAR)

Interface Signal	Source	Target	Type	Content
Execution	App	Seri	Sender / Receiver	The application Mode Manager sends out the execution command to serial signal manager component
Shutdown Request	App	Seri	Sender / Receiver	The application Mode Manager sends out the shutdown request to the serial signal manager component
Filtered Voltage	App	Seri	Sender / Receiver	The application Mode Manager sends out the filtered supply voltage value to serial signal manager component
Reset Request	Seri	App	Sender / Receiver	The serial signal manager component sends out the reset request to application Mode Manager
Error Handling Request	Seri	AUTO	Client / Server	The serial signal manager component sends out the error information to AUTO
Serial Message	Seri	App Diag Cyber Feat AUTO	Sender / Receiver	The serial signal manager sends the serial signal to all application layer components and AUTOSAR
Confidential Information Verification	Seri	Cyber	Client / Server	The serial signal manager sends the confidential serial signal to Cybersecurity component to verify the authentication.
Diagnostic Service Request	Seri	Diag	Client / Server	The serial signal manager sends the diagnostic service request to diagnostic service component, such as the internal information query. internal information query.
Event Information	App Diag Cyber Feat AUTO	Seri	Sender / Receiver	All components send the event information to serial signal manager for the serial communication

Table 3.4-5 Serial Signal Manager Interface

3.4.2.4   *Diagnostic Service*

In the field, all maintenance activities, such as information query, parameter update, software content update, service routine execution, are supported by the Diagnostic Service, which can be done either by using the tester device to connect to the vehicle on-board diagnostic (OBD) connector to communicate with the ECU in the vehicle, or by using the wireless ECU that is connected to the vehicle network to transmit the service requests, the data flow of which and the interactions between the Diagnostic Service component and other components are illustrated in Figure 3.4-15 Diagnostic Service Component Data Flow below:

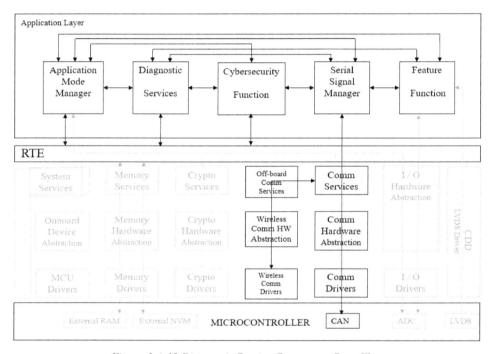

Figure 3.4-15 Diagnostic Service Component Data Flow

The Diagnostic Services implement the Unified Diagnostic Services (UDS) standard also known as ISO-14229, which defines how messages should be formatted and timed for those services listed in Table 3.4-6 UDS Services below, in which, the services in bold font are commonly used.
(Note: the sign of "$" in the service ID represents that the ID is in hex.)

All the automotive ECUs that implement the UDS act as the services' server, and the device connected to a vehicle on-board diagnostic (OBD) connector or the wireless ECU that transfers the diagnostic service requests acts as the services' client, so all the defined diagnostic services are executed in the client-server manner.

Service ID	Description
$10	**Diagnostic session control**
$11	**Request ECU reset**
$14	**Clear DTC**
$19	**Read DTC information**
$22	**Read data by ID**
$23	**Read memory by address**
$24	Read scaling data by ID
$27	**Security access**
$28	**Communication control**
$2A	Read data by periodic ID
$2C	Dynamically define data ID
$2E	**Write data by ID**
$2F	**I/O control by ID**
$31	**Routine control**
$34	**Request download**
$35	Request upload
$36	**Transfer data**
$37	**Request transfer exit**
$38	Request file transfer
$3D	**Write memory by address**
$3E	**Tester present**
$83	Access timing parameter
$84	Secured data transmission
$85	**Control DTC setting**
$86	Response on event
$87	Link control

Table 3.4-6 UDS Services

To ensure the vehicle safety, the vehicle status needs to satisfy predefined prerequisites, and other conditions such as security access check should be passed to execute the required diagnostic services, which is illustrated as Figure 3.4-16 Diagnostic Service Enabler below.

There are three diagnostic sessions defined:
- Default diagnostic session
- Programming diagnostic session
- Extended diagnostic session.

All diagnostic services in the list Table 3.4-6 UDS Services above must be executed in one of those three sessions:

3.4.2 System Structure Design

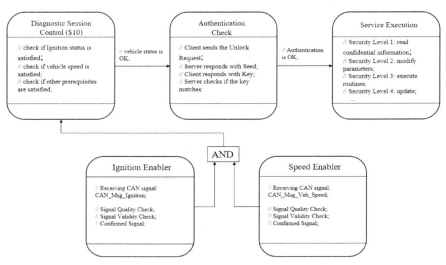

Figure 3.4-16 Diagnostic Service Enabler

Default diagnostic session:

This is the normal ECU operation session, which does not need to use the Diagnostic session control service ($11) to enter. In which, there are only a few of diagnostic services that should be implemented, among which there are only two that are commonly used: Read DTC information service ($19), Read data by ID service ($22) to do the normal maintenance, which will in turn send the information requests to other application layer components and the AUTOSAR components, then transfer the information back to the client.

Programming diagnostic session:

This is the session to update the ECU software and parameter contents, the Diagnostic session control service ($11) is used to enter to enter the session, and the Security access ($27) must be used to clear the authentication check.

Once the ECU enters the session, the diagnostic service component will send the Reset Request to the application mode manager, which will send the Shutdown request with the programming reset parameter to all other components, after which, the ECU will stay in the bootloader control to program the ECU.

The automotive ECU bootloader is a standalone software which implements the Request download service ($34), Transfer data service ($36), Request transfer exit service ($37), and Routine control service ($31) to program the ECU contents: software or parameters, or both.

Extended diagnostic session:

This is the main diagnostic session, in which almost all diagnostic service is implemented except the programming services ($34, $36, $37).

In the session, the specific diagnostic tests can be done by using the I/O control by ID service ($2F), Routine control service ($31), parameters can be modified by the Write data by ID service ($2E), and before those services are invoked, usually the authentication check should be done using the Security access ($27).

The Diagnostic Service component uses the following interfaces to interact with other components:

Diagnostic Service Interface (Legend: App-> Application Mode Manager, Diag->Diagnostic Service, Seri->Serial Signal Manager, Cyber->Cybersecurity, Feat->Feature Function, AUTO-> AUTOSAR)				
Interface Signal	**Source**	**Target**	**Type**	**Content**
Execution	App	Diag	Sender / Receiver	The application Mode Manager sends out the execution command to the diagnostic service component
Shutdown Request	App	Diag	Sender / Receiver	The application Mode Manager sends out the shutdown request to the diagnostic service component
Filtered Voltage	App	Diag	Sender / Receiver	The application Mode Manager sends out the filtered supply voltage value to the diagnostic service component
Reset Request	Diag	App	Sender / Receiver	The diagnostic service component diagnostic service component sends out the reset request to application Mode Manager
Diagnostic Service Request	Diag	App Cyber Seri Feat AUTO	Client / Server	The Diagnostic Service component sends the diagnostic service request to all the application layer component and AUTOSAR to execute the services.
Error Handling Request	Diag	AUTO	Client / Server	The diagnostic service component sends out the error information to AUTOSAR
Confidential Information Request	Diag	Cyber	Client / Server	The diagnostic service component sends out the confidential information request to the cybersecurity function component, such as the ECU ID, security access (0x27).
Event Information	Diag	Seri	Sender / Receiver	The diagnostic service component sends the event information to serial signal manager to communicate with other ECUs, such as the subject ECU status.
Diagnostic Event Notification	AUTO	Diag	Sender / Receiver	The AUTOSAR sends the diagnostic event notification about the error handling to the diagnostic service component to enable the error handling and fault recovery.
Serial Message	Seri	Diag	Sender / Receiver	The serial signal manager sends the serial signal to the diagnostic service component

Table 3.4-7 Diagnostic Service Interface

In some situations, due to the network structure reasons, some ECUs need to implement the functions that transfer the diagnostic service requests and responses between other

ECUs and the diagnostic service tester, which is the definition of diagnostic service gateway, for example, in the left camera BSD system, the BSD should be the diagnostic service gateway between the left BSD camera and the diagnostic service tester. In such case, the ECU design should take the resources and latency impact into consideration.

Nowadays, more and more ECUs are required to be updated using the wireless communication, i.e., the update over the air (OTA) functions, those diagnostic services have the special requirements about the ECU resource, performance, communication integrity check and the update consistence insurances, especially the memory capacity.

3.4.2.5 Cybersecurity Function

The Cybersecurity Function provides the interfaces that are listed in Table 3.4-8 Cybersecurity Function Interface below to other application layer components for the cybersecurity functionalities that are described in the section of 3.4.7 Cybersecurity by cooperating with the Crypto Service Manager (CSM) component in AUTOSAR illustrated in Figure 3.4-17 Cybersecurity Function Data Flow below:

Cybersecurity Function Interface (Legend: App-> Application Mode Manager, Diag->Diagnostic Service, Seri->Serial Signal Manager, Cyber->Cybersecurity, Feat->Feature Function, AUTO-> AUTOSAR)				
Interface Signal	**Source**	**Target**	**Type**	**Content**
Execution	App	Cyber	Sender / Receiver	The application Mode Manager sends out the execution command to cybersecurity function component
Shutdown Request	App	Cyber	Sender / Receiver	The application Mode Manager sends out the shutdown request to the cybersecurity function component
Filtered Voltage	App	Cyber	Sender / Receiver	The application Mode Manager sends out the filtered supply voltage value to cybersecurity function component
Reset Request	Cyber	App	Sender / Receiver	The cybersecurity function component sends out the reset request to application Mode Manager
Error Handling Request	Cyber	AUTO	Client / Server	The cybersecurity function component sends out the error information to AUTO
Diagnostic Event Notification	AUTO	Cyber	Sender / Receiver	The AUTOSAR sends the diagnostic event notification to cybersecurity function

				component, such as the DTC inhibit, DTC recovery.
Diagnostic Service Request	Diag	Cyber	Client / Server	The diagnostic service component sends the diagnostic service request to cybersecurity function component, such as the secure access (0x27).
Confidential Information Verification	Seri	Cyber	Client / Server	The serial signal manager sends the confidential serial signal to Cybersecurity component to verify the authentication.
Cybersecurity Function	Cyber	Diag Seri AUTO	Client / Server	The cybersecurity function sends the security responses to all application layer components, and sends out requests to the AUTOSAR. The commonly used cybersecurity functions are listed in the Table 3.4-22 Common Cybersecurity Function

Table 3.4-8 Cybersecurity Function Interface

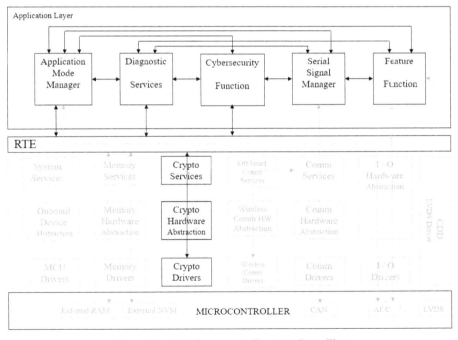

Figure 3.4-17 Cybersecurity Function Data Flow

All the services that are provided by the secure hardware via the crypto service manager are implemented as the service server in the client-server manner by the cybersecurity function component, which can be configured as either synchronous or asynchronous services, the application calls to them are the client requests.

The Cybersecurity Function component needs to check the callers' validity to enhance the security in the application, such as the security challenge using the diagnostic service 0x27, if there are suspicious access requests that are more than the configured access threshold value, the cybersecurity function component should either stop process the request and register the event, or inform the application mode manage to handle.

The interfaces from the Cybersecurity Function component represent the following measures of:

- Authenticity
- Integrity
- Confidentiality
- Privacy

To provide the measures above, the following functions need to be implemented:

- Secure communication, to ensure that the serial communication messages are authenticated and confidential. For which, the Cybersecurity Function component provides the interfaces for encryption and decryption of serial messages.
- Secure diagnostic requests, to ensure that the critical diagnostic services are authenticated, such as reading and modifying the critical information, executing critical services. For which, the Cybersecurity Function component provides the interfaces for authentication check using the diagnostic security access service ($27).
- Secure programming, to ensure that the ECU contents are authenticated and confidential. For which, the Cybersecurity Function component provides the interfaces for authentication check using the diagnostic security access service ($27), and the interfaces for the content's authentication check using the public-private key algorithm.
- Secure Storage, to ensure that the contents stored in the external NVM are authenticated and confidential. For which, the Cybersecurity Function component provides the interfaces for content encryption using the symmetry key algorithm such as AES 128/256.
- Secure startup, to ensure that contents executed are authenticated. For which, the Cybersecurity Function component provides the interfaces for content decryption using the symmetry key algorithm such as AES 128/256.
- Secure debugging interface, to ensure that only the authenticated users can access the debug interfaces, such as JTAG, XCP. For which, the Cybersecurity Function component provides the interfaces for authentication check using the diagnostic security access service ($27).
- Secret information management, such as the encryption keys, random number.
- Intruder protection, to prevent the ECU from the attackers using the devices' features, such as the pins that are exposed externally. Those protections are done by the device drivers, which is out of the application software scope.

3.4.2 System Structure Design

Check List 4 - System Structure Design

- Feature Function
    - o Are all the needed calculations and data in the feature functions defined (data types, data characteristics, data resolutions, data flows, relationships) defined?
    - o Are the definitions above detailed enough to be allocated to the specific devices (for example, some data calculations need the matrix operation ability or the image process ability)?
    - o Are the derivations in the feature functions simulated to prove that they are correct and feasible?
    - o If the special devices are needed by the feature functions, then are those devices feasible to be implemented, and are their drivers (possible done by the CDD) defined?
    - o Are those feature function self-calibrations and self-tests defined?
    - o Are those feature function configurations defined?
    - o Are those feature function startups and shutdowns (which need the cooperation with the Application Mode Manager) defined?
    - o Are the feature function vehicle status enablers defined?
    - o Are the feature function environment status enablers defined?
- Application Mode Manager
    - o Is the application mode manager state machine defined?
    - o Is the cooperation (interfaces, data flows) between the application mode manager and the AUTOSAR defined?
    - o Is the cooperation (interfaces, data flows) between the application mode manager and other application layer components (feature function, serial signal manager, diagnostic service, cybersecurity function) defined?
    - o Is the strategy of startup, shutdown, sleep and wakeup defined?
    - o Are the hardware health monitors defined?
    - o Are the software health monitors defined?
    - o Are the environment health monitors defined?
    - o Is the "fault-safe" strategy (fault states, fault reports and recovery) defined?
    - o Are the vehicle network management strategies including the vehicle network interface state management defined?
- Serial Signal Manager
    - o Are the communication strategies (protocol, receiving and transmitting messages, network management) between the host vehicle and the subject ECU defined?
    - o Is the cooperation (interfaces, data flows) with the AUTOSAR defined?
    - o Is the cooperation (interfaces, data flows) with other application layer components defined?
    - o Are the vehicle communication interfaces states defined?
    - o Are the Integrity and Validity ensuring measures for both receiving and transmitting messages defined?
- Diagnostic Service
    - o Are the required diagnostic services (session, service function, DID, RID, DTC) defined?

- o Are the specific diagnostic services implemented in the dedicated diagnostic session (programming services must in the programming session, ECU test services must in the extended session)?
- o Are the diagnostic service enablers and the specific session prerequisites to ensure the vehicle safety defined?
- o Is the cooperation (interfaces, data flows) with the AUTOSAR defined?
- o Is the cooperation (interfaces, data flows) with other application layer components defined?
- o If the subject ECU is a diagnostic service gateway to other ECUs, do the diagnostic service transfer functions meet the requirements?
- o If the subject ECU should have the update over the air (OTA) functions, do those diagnostic services meet the requirements?
- Cybersecurity Function
  - o Does the cybersecurity function provide the required interfaces of symmetric cryptographic algorithms (AES, DES, CBC, ECB, CFB, CMAC, GMAC)?
  - o Does the cybersecurity function provide the required interfaces of the asymmetric cryptographic algorithms (RSA, ECC)?
  - o Does the cybersecurity function provide the required interfaces of the Secure Hashing Algorithm (SHA)?
  - o Does the cybersecurity function provide the required interfaces of the key management (capacity, key update methods, secret counter)?
  - o Is the cooperation (interfaces, data flows) with the AUTOSAR defined?
  - o Is the cooperation (interfaces, data flows) with other application layer components, especially the secure access interfaces (secure diagnostic access 0x27, JTAG and XCP lock / unlock) defined?

3.4.3 Electronic Architect Design

The goal of electronic architecture design is to provide the needed electronic devices to support the functionalities from the system operation concept based on the data flow, data capacity and control time sequence, which should be based on the estimation and calculation of the operation capacity needed, functional features and system performance, such as:

- What kind of information needs to be processed, such as video image, radar radio, digital signals, environment temperature and power supply voltage analog signals.
- How much information needs to be processed in what time, i.e., what is the system performance requirements.
- What kind of safety requirements need to be satisfied, such as ASIL B or ASIL D.
- What kind of cybersecurity requirements need to be satisfied, such as, secure programming and updating, communication authentication, external memory content confidentiality.

In the electronic architecture design, the electronic elements that construct the ECU illustrated in Figure 2.2-2 ECU Input and Output Signals should be selected to support the next development steps, such as functionality allocation, system safety and cybersecurity design, which needs to consider the specific devices where those functions are allocated.

3.4.3.1 Microcontroller Selection

The most important element in the automotive ECU is the microcontroller, and the typical one is illustrated in Figure 2.2-4 Microcontroller. The following factors should be considered during the selection:

The system performance requirements

Nowadays, the automotive microcontrollers' performance is very high, some of them can run at such high speed like a few hundred million instructions per second, which usually is enough to handle the conventional automotive signals. However, as the new technologies are going into the vehicles, such as autonomous driving, vehicle network, it needs to consider if the selected microcontroller can process the required signals.

For example, in the BSD system, the used camera has the below signals rate:

Resolution: 1620 x 1280 = 2073600
Pixel Depth (BPP): 28 (HDR Support)
Frame Rate: 30 FPS
Blanking: ~20% added

So, the total of signal rate from the camera is: Res x BPP x FPS x 1.2 (blanking) = ~2.1 Gbps.

To accommodate such signal, the ECU input device for the camera signal must therefore handle at least 2.1 Gbps input speed, the CPU (or the microcontroller's core) that calculate the objects should consider the required computing capacity and how much information needs to be transferred between the CPU, the RAM and the external NVM to do such calculation.

The system safety requirements

To check if the microcontroller meets the safety requirements, the following three aspects should be considered:

- The built-in safety mechanism: Every modern microcontroller, especially the automotive microcontroller, provides quite comprehensive built-in safety

mechanisms, such as:
- o Random Hardware Error protection, such as Cyclic Redundancy Check (CRC), Checksum, Error Correction Code (ECC).
- o Memory protection, which enables configuring partition memory into regions and setting individual protection attributes for each region, performing the access permission checks according to the assigned attributes; and provides address translation for an I/O device to identify more than its actual addressing capability to provide the memory isolation, which in turn avoid the I/O devices to corrupt system memory.
- o Peripheral Protection, which provides the isolation of each peripheral, which can control the power for each device individually and isolate each device functions from others.
- o Lockstep running mode, which can meet the ASIL D by using two identical RPU cores in the microcontroller, both can be configured to execute the same instructions at the same time, then compare the result in every step. The microcontroller would report the error status if any of step comparison found any difference. In contrast to that, the other cores, such as APU cores, GPU cores are usually rated as QM devices.
- The safety mechanism that is available from the microcontroller supplier: usually the microcontroller supplier can provide the software solution to enhance the certain devices to meet the higher safety requirements, such as for the APU cores that are originally rated as QM to achieve certain safety rate level like ASIL B in cases where it is needed.
- The safety mechanism that can be developed: if the safety requirements cannot be met in the first two cases, then the ECU developers must develop the necessary solutions to meet the safety requirements.

So, for the safety relevant system, the microcontroller must meet one of below:
- The microcontroller can provide all needed safety functions
- The microcontroller can provide some of needed safety functions, and the rest can be developed by the users based on the APIs from the microcontroller.
- Based on the selected microcontroller and the safety functions from the microcontroller supplier, the needed safety functions can be developed.

The system cybersecurity requirements

The cybersecurity principle is that the "secure world" must not be accessible from the "non-secure world", where the "secure world" is defined as the devices that operate and store the secret information, such as encryption and decryption, authentication and verify authentication, secrete keys, random number, etc., and the "non-secure world" are anything else besides the secure world.

When selecting the microcontroller, the cybersecurity functionalities provided by the built-in devices must meet the required cryptographic algorithm requirements, such as secrete algorithm described in 3.4.7.3 Cryptography: AES, RSA, HAS, key generation and random number generation, secrete information storage: keys and relative information can be stored in the NVM. The commonly used cryptographies in the automotive industry are listed in the Table 3.4-22 Common Cybersecurity Function.

If the built-in devices cannot meet the required cybersecurity requirements, then the microcontroller must allow the modifications or needed development to be done inside of the secure world. However, most microcontrollers don't support such development inside of the built-in security devices, only a few of them do, for example, the microcontrollers that support the special ARM TrustZone structure like Xilinx chips.

3.4.3.2 Power Supply Component Selection

When selecting the power supply component, the first consider factor is the output power capacity, if the ECU is a safety product, then the safety factor should be considered, as well.

Output Power Capacity:

From Figure 2.2-2 ECU Input and Output Signals, the power supply component should provide the power to the microcontroller, the Communication Interface Components, such as the CAN transceiver, ethernet transceiver, serializer / deserializer; the external RAM, the external NVM.

Each device above has different requirement about the power supply' voltage, current, precision and stability, so, the selected power supply component should cover all requirements from all the devices in the ECU.

Power Supply Safety features:

If the system under development is a safety product, then all the devices in the ECU have the essential safety requirement about the power supply to ensure that the input power is in the device's required working range, i.e., the input power voltage should be monitored. Some modern power supply components have the built-in self-monitoring mechanism and output self-regulating measures, so, those features in the power supply component should be considered if needed.

Power Supply Sleep-Wake up features:

If the power supply unit is the wake-up device, then it should meet the wake-up signal requirement, which usually is the case when the ECU is wakened up by the power on signal.

Even if the power supply unit is not the wake-up device, it should be triggered by the wake-up signal from the wake-up device which is usually the serial communication interface component.

3.4.3.3 Communication Interface Component Selection

For all the vehicles, the vehicle designer will provide the communication interface component selection list, in which, there are the vehicle designer approved components for the ECU suppliers to select. In this way, the vehicle designer can make sure that the ECUs in the vehicle will communicate each other which is related to the data transfer rate in the selected communication protocol.

During the selection, the sleep and wake-up features that are supported by the communication interface components should be considered, some of them can support the wake-up by the specific messages without the microcontroller involvement, and some of them are able to have certain functionalities even when the microcontroller is power off.

3.4.3.4 External Memory component (RAM, ROM, NVM) Selection

In most automotive ECU development, the selected microcontroller's internal RAM and ROM capacity cannot meet the system requirements, then the development needs to use the external memory devices to support the system requirements.

The ECU designer should select the external memory components, such as RAM,

ROM, Flash ROM, according to the following considerations:

- If the microcontroller's external device interface matches the memory chip's interface, such as SPI, Quad-SPI, NAND, PCI and ARA.
- If the external memory devices meet the system capacity requirements, usually those memory components' capacity should have about 20% ~ 30 % margin at the end of the development to support the future functionality extension.
- If the external memory devices meet the system performance requirements, which will depend on both the memory devices' capacity and their data transfer speed.
- If the external memory devices meet the required safety requirements. In the case where the safety requirements cannot be met fully by the devices, then the consideration will depend on the ability for the further development to meet those requirements.

3.4.3.5   Environment Interface Device

All the environment information must be converted to either analog or digital signal in order to be processed by a computer, and all automotive microcontrollers have the built-in devices to measure the analog and digital signals. When selecting the microchip, the built-in devices processing speed, resolution and precision should be considered. And in addition to the built-in devices, the special signal convertors or signal sensors should be used to convert the environment signals to the electronic signals, and the relevant external circuits must be designed for the purpose of signal filter and for the built-in device protections. And the environment signals must match the built-in device measure ranges after the signal conversion and signal filtering.

- Analog Signal Input

Currently a computer can only measure one type analog signal: Electronic Voltage, by using the built-in Analog – Digital Convertor (ADC), so all the information that is continuously changed should be handled this way: first, changing the information into the electronic voltage using the special convertor, then the voltage is measured by the computer using the ADC. Such as the measurement of temperature, pressure, liquid level, air flow, etc.

- Analog Signal Output

Currently a computer can only output one type analog signal: Electronic Power, which will be done using the built-in either Digital – Analog Convertor (DAC) or Pulse Width Modulation (PWM) device in the computer, so all the devices that are controlled by the computer will receive such signal to adjust the dynamic status, such as motor motion control, radio volume control, LED lightness control, etc.

- Digital Signal Input and Output

Digital signals, such as switch on or off, LED light on or off, frequency, will be processed by the built-in I/O ports in a microcontroller.

- Special Purpose Signal: either Analog or Digital

The special environment signals that interact with a vehicle must be converted into either electronic voltage or digital signals in order to be processed by a microcontroller. For example, to measure a vehicle speed, the vehicle speed sensors are used to convert the vehicle speed into the digital signals; to measure the objects in the Field of View (FoV), an automotive radar must convert the radio signals into the electronic voltages; and for a camera to measure the objects in the Field of View, it must convert the sense into the digital pixels.

3.4.3 Electronic Architecture Design

Check List 5 - Electronic Architecture Design

- Microcontroller Selection
  - o Does the microcontroller meet the system performance requirements including the performance requirements in both development and production phases?
  - o Does the microcontroller meet the system operation management requirements such as the timer, watchdog, clock, platform management unit (PMU), memory management unit (MMU), peripheral management unit (PMU)?
  - o Does the microcontroller meet the required hardware safety ratings (LFM, SPFM, PMHF) including both the devices and the drivers?
  - o For each safety relevant hardware device in the selected microcontroller that does not meet the required safety rating, does the microcontroller provide the development measures to reach the required safety rating?
  - o For each safety relevant hardware device in the selected microcontroller that does not meet the required safety rating, and the microcontroller does not provide the development measures to reach the required safety rating, are there feasible ways to develop the measures to reach the required safety rating?
  - o Does the microcontroller meet the system cybersecurity requirements including the built-in functions of symmetric cryptographic algorithms, asymmetric cryptographic algorithms, Secure Hashing Algorithm (SHA), the Random Number function (TRNG, PRNG), the key management (capacity, key update methods, secret counter)?
  - o If there is any cryptographic algorithm that the microcontroller does not provide, then can it be implemented based on the isolation between the "Secure World" and the "Non-Secure World" in the microcontroller?
  - o Does the microcontroller meet the system input and output requirements?
- Power Supply Selection
  - o Does the power supply meet the system operation requirements including the capacity, precision, ramp up speed, wakeup by signals, sleep leak current?
  - o Does the power supply meet the system safety requirements including the self-regulation and self-protection, external watchdog?
- Communication Interface Selection
  - o Do the communication Interfaces meet the vehicle communication requirements (capacity, speed, impedance matching, noise resistance)?
  - o Do the communication Interfaces meet the vehicle network sleep and wakeup strategy?
- External Memory Selection
  - o Do the external memory chips (RAM and ROM) meet the system performance requirements including the connectivity, capacity, speed requirements in both development and production phases
- Environment Interface Selection

3.4.3 Electronic Architecture Design

- o Are the analog input and output signals meet the system operation requirements (signals range, resolution, precision, speed)?
- o Are the digital input and output signals meet the system operation requirements (frequency, precision)?
- o Are the special input and output signals meet the system operation requirements (signals range, resolution, precision, speed)?

3.4.4 Functionality Allocation

The purpose of functionality allocation is to allocate each designed function on to the suitable devices to achieve the required functionalities, safety, cybersecurity and system performance. For example, to allocate the functions on to the devices illustrated in Figure 2.2-4 Microcontroller, the following should be considered:

- What functions should be allocated to the RPU?
- What functions should be allocated to the APU?
- What functions should be allocated to the GPU?
- What functions should be allocated to the FPGA?
- Which CPU should process the serial communication interfaces, such as CAN, ethernet?
- Which CPU should process the feature functions' input signal, such as the camera image pixel signal?
- Do those allocations meet the system performance, so that the output signal timing will meet the requirement?

When allocating functions, not only the devices' calculating capacity, but also the devices' properties should be considered. The devices' property is the designed device intention to do certain functions. For example, in Figure 2.2-4 Microcontroller, the RPUs are intended to execute the safety functions, such as the functions rated as ASIL B or C or D; the APUs are intended to execute high speed calculations; the GPUs are intended to execute the graphic processes; the FPGA is intended to execute high speed-required functions; the Secure Hardware Extension (SHE) is intended to execute the cybersecurity functions.

In general, the following facts should be considered:

- Cybersecurity
- Safety
- Latency

3.4.4.1 Cybersecurity

The cybersecurity algorithm functions, such as the AES, RSA, must be executed in the dedicated device that cannot be access by non-security functions. because those cybersecurity functions use the secret information: the encryption keys.

Nowadays, most of automotive microcontrollers have the dedicated built-in security processor that is called Security Hardware Extension (SHE), or Crypto Service Engine (CSE), or Hardware Security Module (HSM), which has its own Arithmetic and Logical Unit (ALU), RAM, ROM and NVM, all of which can be only accessed by the security processor, cannot be access by other cores like RPU, APU, GPU, etc., so, the confidential information inside of the security processor is secured.

The functions provided by the SHE are defined by the chip manufacturer, which cannot be changed by the ECU developers. If those functions cannot meet the development requirements, for example, the SHE provides AES128, however the system under development needs the AES256, or, if the SHE provides the AES256CCM, however the system under development needs the AES256 CMAC, then the development needs to develop the required security functions in the very special way, in which, all the devices for the security functions must meet such requirements: the dedicated security processor, RAM, ROM, NVM cannot be accessed by other non-functions in the microcontroller.

Some microcontrollers based on the ARM TrustZone architecture provide such abilities, but most of controllers don't have such ability.

The commonly required cryptographic algorithms are listed in the Table 3.4-22 Common Cybersecurity Function, which are categorized:

- Symmetric: Advanced Encryption Standard (AES), Data Encryption Standard (DES)
- Asymmetric: Rivest-Shamir-Adleman (RSA) Cryptography, Elliptic Curve Cryptography (ECC).
- Secure Hashing Algorithm (SHA)
- Random Number
- Key management

3.4.4.2   Safety

The first fact to be considered to allocate the safety components is what ASIL level the components need.

The software components that have the ASIL ratings should be allocated to the processors that have the same or higher ASIL ratings, for example, the ASIL B components should be allocated to RPU cores which are rated as ASIL C, and in some microcontrollers rated as ASIL D. If allocating the ASIL rating components to the QM processors, such as APU cores, is needed, then the extra safety measures should be implemented to make sure that the QM devices meet the safety requirements.

The second fact to be considered is the Freedom From Interference (FFI).

If there are different ASIL ratings software components in a processor, then the software components which are developed according to a low ASIL rating may interfere by wrongfully accessing memory regions of software components with a higher ASIL rating, so that the memory related errors in the low ASIL rating software components will propagate to the higher ASIL rating components, which will compromise the higher ASIL rating components' safety.

For example, If both an ASIL B software component and a QM software component exist in the same processor using the shared memory to transfer data, the QM component will lock the shared memory when it modifies the contents, and the lock timing is not ensured because the component is a QM, if the locking memory becomes deadlock, or the locking is longer than design time, then the ASIL B component will not access the memory contents as design. The similar situations can happen in memory stack access, execution switch, etc., which should be avoided by the system design.

According to ISO 262621, if the system software consists of software components with different ASIL ratings, then either the entire software must be developed according to the highest ASIL, or freedom from interference shall be ensured for software components with a higher ASIL rating from elements with a lower ASIL rating. There are two ways to implement the freedom from interference:

- Hardware: allocating the different ASIL rating components to the different processors. For example, if there are the ASIL A and QM ratings software components in the system under development, then the system should allocate all the ASIL A components to one of APU core and allocate all the QM components to another APU core, and those two APU cores don't have the shared memory aera, or the shared memory aera is safely guarded by the designed and implemented safety measures.

▪ Software: This is to divide the memory map into the memory, then allocate the same ASIL rating components to the same partition, and the partitions are managed by the operating system such that the code executing in one partition cannot modify the memory in a different partition, nor can execute the code in a different partition.

If the partitions are managed by AUTOSAR, then structure is illustrated in Figure 3.4-18 Multiple Partitions below, in which:

- Every partition should have its own Basic Software Mode Manager component (BswM) configured specifically to manage the AUTOSAR basic software components in the local partition, such as initialize de-initialize and collaborate the BSW components. the synchronization of the different partition local BswM instances can be accomplished by normal mode-communication (mode request, mode switch) between BswM service.

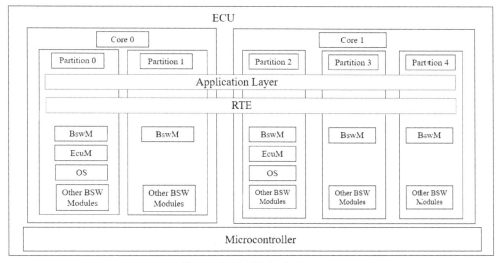

Figure 3.4-18 Multiple Partitions

- Every processing core should have its own ECU State Manager components (EcuM) to manage the processing core about
  - Starting up: the core that started up by the bootloader is the master core, the EcuM in it is the Master EcuM that will start up other EcuMs in other cores. After startup, each EcuM will hand over control to the partition local BswM, which then takes care of the initialization of the other local BSW Modules. Afterwards, the partition local BswM signals the readiness of the partition to the other BswM instances running in other partitions, which is done using normal mode-communication between the BswM service components.
  - Shutting down: the determination will be informed to the partition local BswM to prepare the shutdown by cleaning up, and signals its current state to the other BswM running in other partitions to prepare the collaboration.

- o Restarting up: Based on the need, the EucM can restart the partitions in the core. If a partition is restarted, the local BswM informs to the other BswM instances that it is in a restart mode. Then, the other BswMs can determine if local applications need to be informed or potentially restarted, and how to synchronize them to the newly started partition.
- Every processor (core) should have its own Operation System component (OS) to manage the memory and peripherals, and all processors should have only one Real Time Environment (RTE), which manages the communication between the application and the AUTOSAR components.

The third factor to be considered is the communication between the safety functions and the non-safety functions, or the communications between the functions that have the different ASIL ratings, the data that are transferred from the non-safety to the safety functions or the lower ASIL rating to the higher ASIL rating functions must be guarded to ensure the vale and timing:

- Value check:
  - o CRC, checksum.
  - o Range checks.
  - o Plausibility checks, such as using the experience model, comparing the data from the simulation results.
- Timing check:
  - o If the data are transferred by a communication protocol, then the timing check can be done using the time defined in the protocol.
  - o If the data are transferred by service call or function invoke, then service code or function code should be monitored by the Internal Watchdog or external watchdog.

3.4.4.3 Latency

The system latency is the information processing duration from the time when the input data occur at the system input ports to the time when the output data that react to the input data occur at the system output ports.

The latency of a software function is the duration from the time when the function input data are available to the function to the time when the function output data are available.

The latency of an input hardware device is the duration from the time when the input data occur at the device input ports to the time when the device received data are ready to be used.

The latency design involves the system scheduling, task arrangements, interrupts arrangement, the communication performance and functionality, so, they need to be considered together.

In a processor, there is the only one computing unit: Arithmetic Logic Unit (ALU) that does the computing tasks. How to make use the unit efficiently is done by the scheduling, which is to divide the ALU time into individual pieces that is called as Time Slice, and assign the time slice to the individual tasks, such as the object detection, temperature calculation, CAN message output. So, the scheduling is the critical factor in the system latency.

The ideal case is that all tasks in a processor can be done sequentially in just one scheduler cycle time or one time slice, then the execution will be done cyclically to update

3.4.4 Functional Allocation

the calculated result status.

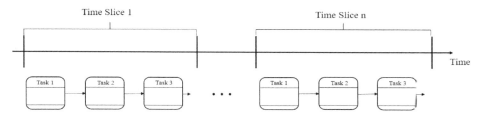

Figure 3.4-19 Sequentially Execution

For example, in Figure 3.4-19 Sequentially Execution above, there ae 3 tasks in a system which can be executed sequentially, and all of which can be done in a time slice, so the execution result will be updated in each time slice.

However, in the reality, some tasks need the data from other functions, such as the data from some peripherals, especially from the asynchronous serial communication peripheral like the CAN interfaces, which will take much more time to have the required data ready comparing with the scheduling time slice. If such tasks are executed in a sequential manner, then those tasks have to wait for the needed data to be ready, during which, the ALU is idle without doing anything, which results in the ALU waste and the whole execution will take longer time. So, the data availability from such functions may not match the scheduled time, in another words, such tasks that need more time to have the needed data ready are not suitable to be executed in such sequential scheduling manner. The common solution is to schedule all the tasks that can be done in the sequential manner in a time slice, and to assign the preparing data tasks to the short time execution functions that usually will be executed in an Interrupt Service Routine (ISR) with a higher execution priority, so that the ISRs will interrupt the sequential execution (preempt) whenever the preparing data tasks need to be handled, in such way, the system can execute the majority calculations in sequential manner, and the data preparing tasks will be done by the ISRs in the preemptive manner.

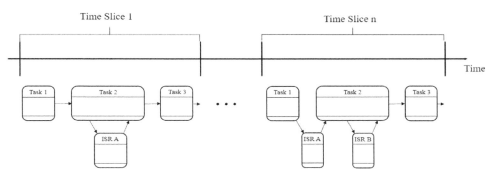

Figure 3.4-20 Execution with ISR

For example, in Figure 3.4-20 Execution with ISR, the sequential execution of task1, task2 and task 3 can be interrupted by the ISR A and the ISR B, and obviously the ISR A and ISR B have the higher execution priority than task 1, task2 and task 3. And such ISRs, as mentioned, usually handle the short execution tasks, such as serial communication peripheral data handling, timer reset, Direct Memory Access (DMA), etc., so that the ALU in a processor will be used efficiently.

The execution in Figure 3.4-19 Sequentially Execution is called as non-preemptive scheduling, and the execution in Figure 3.4-20 Execution with ISR is called as preemptive scheduling, and the preemptive scheduling is commonly used in an automotive ECU development, which involves multiple time factors in the system scheduling, which will impact the latency calculation for the processor.

Latency Calculation:

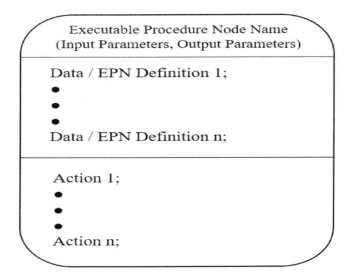

Figure 3.4-21 Executable Procedure Node

The latency for the EPN illustrated in Figure 3.4-21 Executable Procedure Node above is calculated as the time summary of action 1, …, action n in it if there is not any interruption.

The latency for the output data 1 below is calculated as the time summary of input data 1, …, input data 1i, middle data 11, …, middle data 1j plus the $f1$ operation duration if there is not any interruption.

Output Data $1 = f1$ (Input Data 11, …, Input Data 1i, Middle Data 11, …, Middle Data 1j);

The latency calculations above don't consider any interruption, which is not realistic, i.e., in the reality, most systems must handle the interruptions, such as the interruption from the high priority functions, waiting for response from other functions, waiting for the input data from the input device functions.

In a computing system, there are two types of interruptions involved in the latency calculations:

- Waiting for the needed data to be ready, i.e., the needed data is transferred from other functions.
- Interruption from high priority functions.

Data transfer latency calculation:
There are only two types of data transfer in a computing system:
- Shared memory.
- From one location to another controlled by a transfer or communication protocol.

The data transfer can happen between:
- Function and function, which is kind of shared memory data transfer.
- Function and peripheral device, which is controlled by a communication protocol.
- Partition and partition, which is kind of shared memory data transfer.
- Processor and processor, which is either the shared memory data transfer, or kind of communication controlled by a protocol.

The data transfers above are divided into the following two types, and the data transfer latency of those two types data transfer can be calculated:
- Synchronous data transfer, which has the following types:
  o Synchronous Function call with parameters: the transferred data will be available immediately to the receiving function, so the latency is zero.
  o Synchronous communication (either serial or parallel), the data availability time that is the data transfer latency can be calculated based on clock information in the protocol.
- Asynchronous data transfer, which has the following types
  o Asynchronous function call: the latency should be calculated as the timeout value.
  o Asynchronous memory sharing, i.e., the contents availability in the shared memory must be known by periodic polling, whose latency is the timeout value.
  o Asynchronous communication: the latency is the periodic interval value.

Interruption latency calculation:
Interrupt latency = Interruption function duration X interruption frequency.

Latency Impact factors:
There are the following important latency impact factors in a computing system:
- Serial communication signals from a peripheral device, such as CAN, the latency is the signal's periodic interval value.
- Asynchronous client-server function call and the asynchronous shared memory, the latency is the timeout value.

The reasons why those factors impact the system latency will be explained in the BSD example below.

System Schedule
System scheduling should be done by cooperating with the system scheduler manager (SchM in AUTOSAR) to assign the processor execution time to the tasks to execute the design functions, to do which, the following should be considered:
- If the output data are required to be updated at certain rate, then system scheduling should follow the update rate.
- If there are the key input data that are updated at the certain rate, then the system scheduling should be scheduled to make use the input data in time, such as the

BSD example, the input image rate is 30 FPS, i.e., one frame image / 33 ms, so the BSD system scheduling should make use the input image in time, i.e., it is better to schedule the processor(s) at the 33 ms at the main schedulers' interval.

- If the tasks need to be executed sequentially without any interruption, then they should be scheduled in one piece of schedule time slice.
- If the tasks can be executed parallelly, then they can be scheduled in different schedulers in different processors.
- If a data transfer task needs to execute for a long time, then it is better to make it an asynchronous server task

BSD Signal Latency

Taking the BSD system as an example to design the system latency.

The data flow and time sequence of the whole BSD system in the vehicle including the electronic power steering (EPS) control is illustrated as Figure 3.4-22 BSD System Latency below, in which, the T0, T2 and T4 are the sub-systems of object sensing camera, camera BSD and electronic power steering (EPS) control ECUs' latency, and the T1, T3 and T5 are the data transfer latency, and all of them must happen in serial, so there is not latency design option about this aspect.

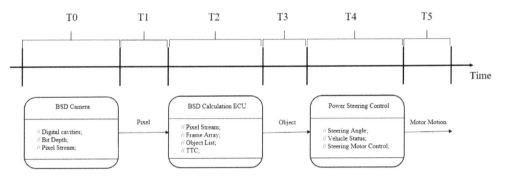

Figure 3.4-22 BSD System Latency

There are two latency design options regarding to the camera BSD system latency design:

- Depending on the output data requirement
- Depending on the input data requirement

To the BSD output data, there is not the specific requirement about either what update frequency should be or what BSD latency should be. The only timing requirement is the TTC = 2.5 second, i.e., the whole BSD system should at least output the detect object information to the electronic power steering (EPS) control for 2.5 seconds before the vehicle makes left turn, which means:

- The whole BSD system should detect the object at the 69.4 meters far if the relative speed is 100 KM/H.
- The whole BSD system should detect the object at the 20.8 meters far if the relative speed is 30 KM/H.
- The whole BSD system should detect the object at the 6.9 meters far if the relative speed is 10 KM/H.

So, the output data timing does not impact the BSD system latency much.

Then, considering the input data, the input data timing will impact the BSD system latency a lot, because the camera pixel signals are transferred at the 30 FPS from the camera to the left camera BSD, which has the following time constraints to the system:

- To the camera BSD system, the input data availability rate is 33 ms
- The camera BSD system scheduling should match the input data rate, if the BSD system scheduler is faster than the input data rate, then system will not have the fresh input data; if the BSD system scheduler is lower than the input data rate, then some input data will be wasted.

So, in the camera BSD ECU, all the object detection processors are scheduled at 33 ms as the system schedulers, and the system will take 4 frames to detect the objects:

- The GPU is scheduled at 33 ms to process the input camera pixel stream signals including the intrinsic data correction.
- The APU0 is scheduled at 33 ms to process the camera image frame, and it will process two frames for 66 ms to prepare the data for the APU1.
- The APU1 is scheduled at 33 ms to track the objects, and it will process the two frames for 66 ms to track the objects and calculate the TTC.

Another factor to impact the system latency is the time to synchronize the processor's schedule with the input data rate, which is: the time when the input data are ready may or may not match the processor scheduler start point, the worst case is that when the input data are ready, the processor just started already, then the input data must wait for the next processor cycle to be processed.

The issue above can be solved by the synchronization between the input device and the processor, i.e., the input data ready status triggers the processor execution. The disadvantage of which is that the development will not be flexible, because the relative components are tight coupled, then if one of them has issue that cause the fault state, then all other components need to handle such issue, as well. So, in the camera BSD system, such tight couple mechanism is not used, rather, the independent coupling mechanism is used below.

Based on the analysis above, the BSD example signal latency is illustrated in Figure 3.4-23 BSD ECU Signal Latency below.

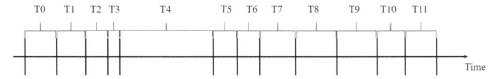

Figure 3.4-23 BSD ECU Signal Latency

The signals in the BSD system are listed in Table 3.4-9 BSD Signal Latency below including the signals in the three ECUs involved in the BSD example: the object detection sensor (camera), the camera BSD ECU that receives the camera pixel signals from the camera to calculate the object detected, the electronic power steering (EPS) control ECU. The signals are:

- T0 is the camera ECU latency, T1 is the camera pixel signal transmission rate that is 30 FPS based on the LVDS protocol, i.e., each frame needs 33 ms latency to transmit from the camera to the BSD ECU, based on which, the camera latency: T0 is scheduled as 33 ms, as well, so that the efficiency of the camera is scheduled at right rate. In the BSD example system, the input signal transmission rate (33 ms) is critical for the system, which decides the object detection calculation processor scheduling interval and the BSD ECU output signal rate, as well, which will be described next.

- T4 is total of processing time for the GPU, APU0 and APU1, which includes the 4 frames used to detect the objects and the TTC calculation.
- T2 and T5 are the delays between the input and output data ready and the object processors' schedulers as analyzed above. And the same for the T7.
- The T6 is scheduled at 33 ms as well to match the processors' scheduling, so the object calculation result can be output efficiently.

Time Interval Name	Time (ms)
T0: Object detection sensor	33
T1: Signal transferred from sensor to the object calculation unit (BSD)	33
T2: BSD input buffer ready	33
T3: BSD synchronize enabler signals	1
T4: BSD object calculation	165
T5: BSD output Buffer ready	33
T6: BSD signal transferred to the electronic power steering (EPS)	33
T7: EPS input buffer ready	33
T8: EPS Synchronize enabler signals	40
T9: EPS driving signal calculation	40
T10: EPS output buffer ready	40
T11: EPS signal output	40
Total:	**524**

Table 3.4-9 BSD Signal Latency

The latency in the electronic power steering (EPS) control ECU is similar to the BSD system.

The total of whole BSD system latency is 524 ms, which will lead to 14.56 meters for a vehicle to travel at the 100 KM/S before taking the actions, i.e., the vehicle will be starting to take an action at 14.56 meters after detecting the object.

Usually, the time-consuming activities in the automotive ECU system are:
- Serial communication: such as LVDS, CAN, Ethernet, FlexRay, in those serial communication protocols, the information is transmitted by the message packages, which will be received then re-assembled at receiver, then the receiver will generate interrupts to inform the host processor that the input signals are ready.
- Asynchronous data transfer, such as the camera ECU input pixel signals delay from ready to the signals to be processed.
- The communication between the processors: such as the communication between the RPU and APU, between the RPU and SHE, which are generally the serial communication, such as I2C, UART.

3.4.4 Functional Allocation

- The communication between the partitions, which usually uses the shared memory, and it will cost time to signal the contents' availability.
- Functional component switch: Different functional components are allocated on different tasks in the different processors, such as: the AUTOSAR and enabler components are allocated on the RPU, the image processing components are allocated on the APU, the cybersecurity components are allocated on the Secure Hardware Extension (SHE), they provide the input information one for another, and the calculation logic needs to synchronize them to follow the relationships defined in the operation concept, i.e., the execution will switch between the tasks and between the processors, which will take time.

So, the requirement of the system latency performance conflicts with the system development strategy:

- From the system development strategy point of view: to simplify the complex system development, the modularization is used, i.e., by separating the complex system into dedicated modules based on the functionality category, the development can be simplified, such as AUTOSAR, the Application Mode Manger, Diagnostic Services, Serial Signal Manger, Cybersecurity Function components in Figure 3.4-7 ECU Component Structure.
- From the system latency performance requirement: the best case is to have all functionalities in just one task executed in one processor, which is obviously not feasible. Then to achieve the better latency performance, the approach is to reduce the separation, which can be achieved by carefully allocating the functional components on to the processors using the optimized allocation strategy:
  - Increase parallel execution as much as possible.
  - Decrease the asynchronous execution and the serial communication as much as possible.

3.4.4 Functional Allocation

3.4.4 Functional Allocation

Check List 6 - Functionality Allocation

- The cybersecurity allocation
  - o Do the symmetric cryptographic algorithms (AES, DES, CBC, ECB, CFB, CMAC, GMAC) from the microcontroller meet the system requirement?
  - o Do the asymmetric cryptographic algorithms (RSA, ECC) from the microcontroller meet the system requirement?
  - o Do the Secure Hashing Algorithm (SHA) from the microcontroller meet the system requirement?
  - o Does the Random Number function (TRNG, PRNG) from the microcontroller meet the system requirement?
  - o Does the key management (capacity, key update methods, secret counter) from the microcontroller meet the system requirement?
  - o If there is any cryptographic algorithm that the microcontroller cannot provide, then can it be implemented based on the isolation between the "Secure World" and the "Non-Secure World"?
- Safety allocation
  - o Do all the safety relevant hardware devices meet the allocated safety ratings (LFM, SPFM, PMHF) from the allocated functional requirements?
  - o For each safety relevant hardware device that does not meet the allocated safety rating, are the corresponding safety enhancements to reach the required safety rating designed?
  - o If the software components with different ASIL rating exist in a processor, is the Freedom From Interference (FFI) mechanism implemented?
  - o If the communications exist between the software components with different ASIL rating in the ECU, is the Freedom From Interference (FFI) protection mechanism implemented in the higher ASIL rating software component that receives signals from the lower ASIL rating software component?
  - o Are the startup, shutdown and operation management strategies defined for each partition, processor?
- Latency
  - o Is the required output signal timing satisfied?
  - o Are the time requirements of power on and off, sleep and wakeup including the initial communications satisfied?
  - o Are the communication strategies defined between function and function, function and peripheral device, partition and partition, processor and processor?
  - o For the synchronous shared memory, is the data transfer mechanism (functional call, semaphore, interrupt) defined
  - o For the synchronous communication, is the data transfer clock defined?
  - o For the asynchronous memory data transfer, is the timeout value defined?
  - o For the asynchronous communication, is the periodic interval value defined?
  - o For each interruption function, are the worst interruption function duration and the maximum interruption frequency defined?

3.4.4 Functional Allocation

- o Does the scheduler comply with the critical input signal update rating if that signal will trigger the output signal update?
- o Does the scheduler comply with the required output signal update rating if the output signals are required to be updated at the certain period?
- o If the preemptive scheduling is used, is the priority strategy defined?
- o Are the tasks arranged such that maximizes the sequential execution?
- o Does the scheduler design consider the operation overheads?

3.4.5   FMEA

The Failure Mode and Effect Analysis (FMEA) is an approach to figure out what could go wrong in the system under development, then based on the outcomes from the FMEA, the measures to prevent those potential failure modes from violating the system functionalities will be considered in the system design to make the system more reliable.

What could go wrong means that there may be some potential errors (which are called as **failure modes** in the FMEA terminology) in the system, and those errors will cause some consequences to impact the system functionalities, the consequences are called as the **effect** to the system in the FMEA terminology.

The errors are divided into two types:

- Human error, which is the mistakes made by the developers, such as the wrong parameter values, using the non-initialized devices or services, wrong configurations. This kind of errors are mainly handled by the quality control, not by the FMEA. However, certain non-foreseeable results, such as the result data derived from complex calculation may be out of range, should be handled by the FMEA.

- Systematic error, which is the system nature, because the computing system is made of millions electronic elements like transistor, and those electronic elements will be aged, which will lead to wrong execution results. Another reason why a computing system may run wrong is due to the electromagnetic interference in a vehicle. Those non-foreseeable and random errors will be handled by the FMEA.

As mentioned in the section of 2.2.3.1 Reliability, in a computing system, all the information is represented by the combination of binary "0" or "1" that represented by the electronic signals. For such system, the systematic error has only two type errors: Value Error and Time Error.

Value error: the binary combination in a computing system can represent either a data or an execution instruction. The value error means that the binary combination has wrong value, which is caused by either the binary combination transmission or transforming:

- The transmission error is caused by transmitting device, such as that a binary combination is transferred from the external flash memory to the internal RAM, and during the transferring, some bits in the content may be changed, such as: "11101100" is transferred from location A to location B where the binary combination becomes "11101101" due to the potential transmission errors that may be caused either by the transmission equipment or the external interference.
  The errors may happen to both data and execution instruction.
    o   If it happens to the data, then the result will be wrong.
    o   If it happens to the execution instruction, then the effect will vary:
      ▪ If the instruction operates some data to derive the result data, then the result data may be wrong, which causes the data transforming errors below.
      ▪ If the execution controls the execution, then the execution may be out of control, such as to result in:
        • Dead lock.
        • Access the contents that should not be accessed.
        • Execute the instructions that should be executed.
      The final consequence of above is that the instructions to

derive the data will not be executed as designed, which will result in the wrong final result data.

- The binary combination transforming error is that the wrong result binary combination is derived from wrong instructions, because the instructions may be changed during the transmission between the memory and the arithmetic logic unit (ALU) in a processor.

Timing error: the binary combination does not occur at the required time. The detection to such error is divided into two parts:

- the contents inside of the microcontroller
- the contents from outside of the microcontroller.

To the contents inside of the microcontroller, this type errors cannot be detected directly, neither by the microcontroller nor by the application, because all the execution instructions in a processing core are executed in serial sequence by the arithmetic logic unit (ALU) based on the system clock, so the detection instruction and the under detected instruction or data run on different time, i.e., they don't have the common referenced time base to check the timing. However, some timing relative measure can be implemented to monitor the execution timing:

- Internal or external watchdog monitoring
- Window watchdog monitoring

To the contents input from outside of the microcontroller, the time errors can be detected based on the input communication protocol's clock information:

- For the synchronized communication, the synchronized clock information should be monitored.
- For the asynchronized communication, the synchronization information is embedded into the protocol, which can be used to monitor the data timing.

FMEA is to analyze the potential errors in:

- the components that make up the system, or
- the development middle results that are derived from the development

After the potential errors are analyzed, the developers will design the protection measures against those errors.

Then the key question is:

- which components or which middle results should be analyzed?
- Or, what granularity should be taken to select the components or middle results that will be analyzed using the FMEA?

The ultimate granularity for software is to analyze every software sentence in the source code.

The ultimate granularity for hardware is to analyze every component in the product, such as every transistor, resistor, capacitor, connector.

However, the ultimate granularity is not feasible. So, selecting the suitable components or middles results to analyze the potential errors is critical to the successful FMEA.

The Data Driven FMEA in this book provides the clearly defined efficient, persistent and optimized approach, which much simplifies the conventional FMEA process.

Note: In most of documents about FMEA, Automotive Safety and in ISO 26262, the terms of Fault, Error, Failure and Failure Mode are distinguished, one of the examples is from the Part 9 of ISO 26262 which is described in the section of 3.4.6.4 ISO 26262 Compliance.

However, in this book, all those terms have the exact same meaning that something is wrong, which means: "The result does not follow the requirement(s)". And the distinguishing them does not add any value to the development, and the errors at the different level can be easily identified by the context, so, in this book, all those terms: Fault, Error, Failure and failure mode are used equally to described something wrong without any other specific meaning.

3.4.5.1 Conventional FMEA

The commonly used Failure Mode and Effect Analysis (FMEA) method in automotive industry is specified in the "Potential Failure Mode and Effects Analysis (FMEA)" (ISBN: 978-1-60534-136-1), in which the Approach section specifies: "There is no single or unique process for FMEA development". The similar one is the FMEA method from Quality-One International. Neither of FMEA methods have the defined explicit and complete approach to do the FMEA, instead of having very generic processes, basically the FMEA developers need to develop the system requirement and design specifications, then to develop the failure modes, failure causes and effects based on the specifications. Such development is totally based on the developers' experience, and highly depended on the specifications which are prone to ambiguous, incomplete and inconsistent, so the result is not certain, and the development is not efficient.

The other commonly used FMEA method is to use the tool of APIS IQ-RM comprising:

Step 1: Develop system structure.

Step 2: Design system functions based on the system structure.

Step 3: Develop function nets based on the system functions.

Step 4: Identify potential failure modes for each the system function.

Step 5: Develop failure nets based on the failure modes and the function nets.

Step 6: Classify each failure mode in the failure nets that is comprises:

classify each failure's severity from 1 to 10, and

classify each failure's occurrence from 1 to 10, and

classify each failure's detection from 1 to 10.

Step 7: Multiply those 3 classification numbers to derive failure mode risk rating for each the failure mode, if the product result is higher, then the failure mode is riskier.

Step 8: Analyze failure mode's effects to the functions based on the failure mode risk ratings, function nets and failure nets.

The issues for the method above are:

- The critical steps in the FMEA are the development of system structure in the step 1 and system functions in the step 2, and there is not clearly defined explicit and complete approach to do so currently. If there is anything that is not accurate or not necessary in the development, then the derived activities from which will not be accurate or not necessary in the following steps.

- The function nets in the step 3 and the failure nets in the step 5 will impact the effect analysis in the step 8, and there is not clearly defined explicit and complete approach to do those steps. If the nets are incorrect, then the effect analysis results are incorrect.

- The failure mode classification definition in the step 6 is vague and it not standardized, and the classification activities are redundant in the safety system development.

Taking the BSD system FMEA as an example, the convention FMEA method is described below.

The first step is to develop the high-level BSD function component structure by designing all the BSD relevant functions in the different levels in the vehicle, then connecting them according to their physical connection relationship in the vehicle, which is illustrated in Figure 3.4-24 BSD Function Structure below.

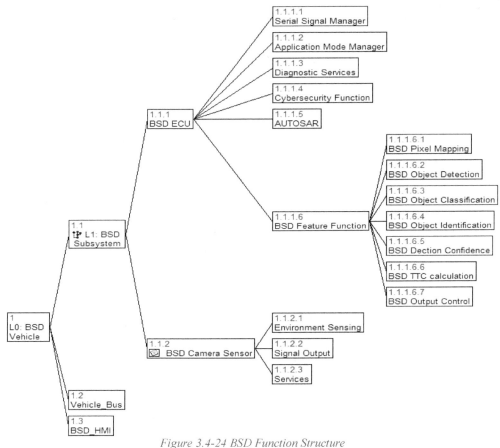

Figure 3.4-24 BSD Function Structure

The second step is to design the system functions based on the system structure, which consists of:

- The top level is the Blind Spot Detection (BSD) function in the vehicle.
- The second level from the vehicle point of view has the 3 components:
  o The vehicle communication bus, via which, the BSD ECU will communicate with other ECUs in the vehicle.
  o The HMI, which is the interface that the drive will interact with the BSD system.
  o The BSD System, which consists of (1) BSD camera (2) BSD ECU that

consists of the components described in the section of 3.4.2 System Structure Design, which are:

- Serial Signal Manager
- Application Mode Manager
- Diagnostic Services
- Cybersecurity Function
- AUTOSAR
- Feature Function, which, in turn, consists of:
  - Pixel Mapping
  - Object Detection
  - Object Classification
  - Object Identification (ID, Label)
  - Object Detection Confidence
  - Object TTC Calculation
  - BSD Output Control

- The BSD camera functions are considered in the FMEA because their failure mode will impact the BSD system, which are:
  - Environment Sensing
  - Signal Output
  - Service

The functions at the most left side in Figure 3.4-24 BSD Function Structure above are the highest-level BSD functions in the vehicle, the functions at the rightest side are the lowest ones.

The lower levels functions are the depositions of the high-level function, i.e., the lower-level functions have the "deriving from" relationships with the high-level functions, and the high-level functions have the "made from" dependency relationship with the low-level function, so, if there are some errors in the lower functions, then those errors will impact the related high-level functions.

Some relationships exist in certain functions which are not connected directly, but there is the functional dependency between them, such as the function: "BSD Output Control" is one of feature functions that depends on the function: "Serial Signal Manager" because the BSD output signals: CAN_BSD_Left_Alert (TTC) are CAN signals that are managed by the Serial Signal Manager, though those two functions are not connected directly. If the Serial Signal Manager function has errors, then the errors will impact the BSD Output Control function, which in turn will impact the BSD ECU, which in turn will impact the BSD sub-system.

The third step in the BSD FMEA development is to connect all level functions by following the relationships between them, either composition - decomposition or functional dependency, which forms the BSD function net, a part of which is illustrated in Figure 3.4-25 BSD Function Net below.

The fourth step is to identify all the potential errors for each function in the function net. For example, there is the function: "The Serial Signal Manager shall send and receive the CAN messages from the Vehicle CAN bus at the 500 KM / S rate" in the Serial Signal Manager which is allocated at the most right-top corner in Figure 3.4-25 BSD Function

Net, for which, there are the following potential errors:
- The CAN interface cannot receive CAN message
- The CAN interface cannot send CAN message
- The CAN interface receives CAN message at wrong time
- The CAN interface sends CAN message at wrong time

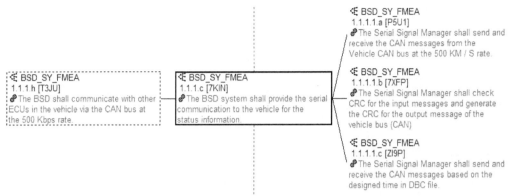

Figure 3.4-25 BSD Function Net

The fifth step is to develop the failure net. In the BSD ECU that is one level higher than the Serial Signal Manager in Figure 3.4-25 BSD Function Net, there is a function: "The BSD system shall provide the serial communication to the vehicle for the status information" which depends the Serial Signal Manager, and there is the potential error in the BSD ECU: "The serial communication interface does not work correctly to follow the requirements" which has the decomposition relationship with the 4 errors listed above. The partial failure net is formed by connecting the 4 errors to the one in the BSD ECU illustrated in Figure 3.4-26 BSD Failure Net.

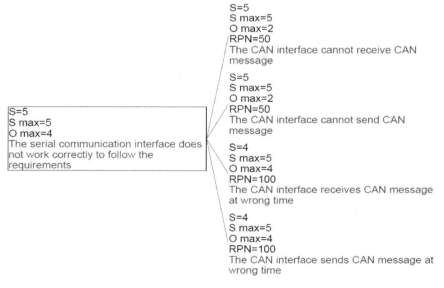

Figure 3.4-26 BSD Failure Net

The sixth step is to classify each failure mode in the failure nets. Once the failure net is established, each failure will be analyzed by assigning the Severity (S), Occurrence (O) and the Detection (D) ratings from 1 to 10:'

Severity:
- 10: very significant
- 9: very significant
- 8: significant
- 7: significant
- 6: medium
- 5: medium
- 4: medium
- 3: small
- 2: small
- 1: not perceptible

Occurrence:
- 10: high
- 9: high
- 8: medium
- 7: medium
- 6: small
- 5: small
- 4: small
- 3: very small
- 2: very small
- 1: improbable

Detection:
- 10: improbable
- 9: very small
- 8: small
- 7: small
- 6: small
- 5: medium
- 4: medium
- 3: medium
- 2: probable
- 1: very probable

The seventh step is to multiply those 3 classification numbers to derive failure mode risk rating for each failure: Severity (S) × Occurrence (O) × Detection (D), the result will be taken as the Risk Priority Number (RPN) that is used to identify which failures have the higher risk: higher RPN, higher risk. For the BSD example, some RPN numbers are listed in.

No.	RPN	S	O	D	Effects	Cause	Preventive action	Detection action
51	50	5	2	5	The serial communication interface does not work correctly to follow the requirements	The CAN interface cannot send CAN message due to HW error	HW Design Review (Requirement & Architecture)	HW Test Case H1
52	50	5	2	5	The serial communication interface does not work correctly to follow the requirements	The CAN interface cannot receive CAN message due to HW error	HW Design Review (Requirement & Architecture)	HW Test Case H2
53	100	5	4	5	The serial communication interface does not work correctly to follow the requirements	The CAN interface receives CAN message at wrong time due to the CAN driver timing design	HW Design Review (Requirement & Architecture)	HW Test Case H3
54	100	5	4	5	The serial communication interface does not work correctly to follow the requirements	The CAN interface sends CAN message at wrong time due to the CAN driver timing design	HW Design Review (Requirement & Architecture)	HW Test Case H4
55	90	5	6	3	The signal values in the serial communication interface are incorrect based on the requirements	The integrity check is not implemented (CRC, check sum)	System Design Review (Requirement & Architecture)	System Test Case S1

					The signal values in the serial communication interface are out of range based on the requirements	The integrity check is implemented wrongly (range check).	System Design Review (Requirement & Architecture)	System Test Case S2
56	90	5	6	3				
57	90	5	6	3	The signal timing in the serial communication interface does not follow the requirements	The validity check is not implemented (time out check)	System Design Review (Requirement & Architecture)	System Test Case S3

Table 3.4-10 Risk Analysis

In the Risk Analysis table above, all the items have the same severity ratings, because all of them impact the CAN serial communication interface. Among them:

- the item No. 53 and item No. 54 have the higher RPN, because those two items failure modes are from the CAN driver design which are difficult to detect, so those two items need most attention among the items in the table.
- The item No. 55, item No. 56 and item No. 57 failure modes are from the application layer, it is easy to detect those errors though their occurrence ratings are higher than others, so their RPN numbers are not the highest.
- The item No. 51 and item No. 52 have the lowest occurrence ratings and they are easy to be detected which are most possibly detected by the CAN Bus Off detection.

The eighth step is to analyze failure mode's effects to the functions based on the failure mode risk ratings, function nets and failure nets, the development can focus on the failures that have higher risks and develop the protection measure against them.

As the development is progressing, the Detection rating can be reduced if the implemented detection measures are working as designed, and the Occurrence rating can be reduced if the protection measures are working, so that the RPN value can be reduced, i.e., the failure's risk is lowered by the implemented measures.

What risk level or what RPN is tolerable to the system, that is the project specific.

3.4.5.2 Data Driven FMEA

In data driven system engineering approach, the FMEA will make use the system operation concept to find the failure modes, failure causes, effects analysis, and risks analysis. Taking the relationship below as an example:

Output Data 1 = $f1$ (Input Data 11, …, Input Data 1i, Middle Data 11, …, Middle Data 1j)

by which, the relationship between the output data 1, input data and meddle result is defined by the operation of $f1$, then the failure modes, failure causes and failure effects of the data will impact each other according to $f1$, in this way, FMEA development logic is exactly the same as the system operation concept as it is supposed to be, so that it can be accurate and efficient.

Another improvement in the data driven FMEA is the risk analysis consisting of assigning a severity level, a probability level and a controllability level for each failure mode in the system under development, in data driven FMEA, the risk analysis is carried over from ISO 26262, in another words, the classifications of severity, probability and controllability are exact same as the ones in ISO 2626, and those activities are mandatory in the automotive safety system development, and they are clearly defined and standardized in the standards, so it will be efficient to re-use those concepts.

In the data driven system engineering, the system operation concept based on the output data, the input data and the middle data can be defined in detail using formulas below which is described in the section of 3.1.1 Data Driven Development:

Output Data 1 = $f1$ (Input Data 11, …, Input Data 1i, Middle Data 11, …, Middle Data 1j);
Output Data 2 = $f2$ (Input Data 21, …, Input Data 2l, Middle Data 21, …, Middle Data 2p);

…

Output Data n= fn (Input Data n1, …, Input Data nq, Middle Data 1n, …, Middle Data nr).

The goal of FMEA is to find all failure modes, all failure causes and failure effects, then to analyze risks based on the definitions above.

The failure modes consist of the failure modes of the Output Data 1 consisting of the intrinsic failure modes of the Output Data 1 and the failure modes from each operated data in the calculation represented by the formula of $f1$ consisting of Input Data 11, …, Input Data 1i, Middle Data 11, …, Middle Data 1j, wherein the intrinsic failure modes of the Output Data 1 are defined as that the Output Data 1 does not behave as implemented; and using the same way to find the failure modes of the Output Data 2, …, Output Data n.

The failure causes consist of the failure causes for each failure mode of the Output Data 1 consisting of the intrinsic failure causes for each failure mode of the Output Data 1 and the failure causes for each failure mode of each operated data in the calculation represented by the formula of $f1$ consisting of Input Data 11, …, Input Data 1i, Middle Data 11, …, Middle Data 1j, wherein the intrinsic failure causes for each failure mode of the Output Data 1 are defined as the reasons that cause each failure mode of the Output Data 1 that does not behave the Output Data 1 as implemented; and using the same way to find the failure causes for each failure modes of the Output Data 2, …, Output Data n.

The failure effects consist of the failure effects for the Input Data 11 consisting of the intrinsic failure effects for the Input Data 11 and the failure effects for each derived data in the calculation represented by the formula of $f1$ consisting only of the Output Data 1, wherein the intrinsic failure effects for the Input Data 11 are defined as the failure effects from each failure mode of the Input Data 11; and using the same way to find the failure effects of the Input Data 21, …, the Input Data 2l, the Input Data n1, …, the Input Data nq, the Middle Data 21, …, the Middle Data 2p, the Middle Data n1, …, the Middle Data nr.

The failure effects consist of the failure effects for the Input Data 11 consisting of the intrinsic failure effects for the Input Data 11 and the failure effects for each derived data in the calculation represented by the formula of $f1$ consisting only of the Output Data 1, wherein the intrinsic failure effects for the Input Data 11 are defined as the failure effects from each failure mode of the Input Data 11; and using the same way to find the failure effects of the Input Data 21, …, the Input Data 2l, the Input Data n1, …, the Input Data nq, the Middle Data 21, …, the Middle Data 2p, the Middle Data n1, …, the Middle Data nr.

The failure effects from the Input Data 11 to the Output Data 1 are same as the failure modes from the Output Data 1 that are caused by the failure modes from one of the operated data in $f1$ that is the Input Data 11, which are same for all pairs between the Input Data and Output Data linked by their calculations, and all pairs between Middle Data and Output Data linked by their calculations.

The analyzing risks consists of assigning a severity level, a probability level and a controllability level for each failure mode in the system under development, wherein classifications of the severity, the probability and the controllability are carried over from ISO 26262; prioritizing the risks according to the multiplication product of severity level, probability level and controllability level of each failure mode, the higher, the risker.

The FMEA method above can be done recursively to any data that need to be decomposed further into decompositions as the development progresses. For example, if the Middle Data 11 needs to be decomposed into such expression: Middle Data $11 = fm11$ (Input Data 111, … Input Data 11i, Middle Data 111, …, Middle Data 11j), wherein the $fm11$ is the calculation to derive the Middle data 11, the input data group of Input Data 111, …, Input Data 11i is a subset of input data group of Input Data 11, …, Input Data 1i, the middle data group of Middle Data 111, …, Middle Data 11j is a subset of middle data group of Middle Data 11, …, Middle Data 1j. Then the FMEA method for the Middle Data 11 will be done by applying the FMEA processes above to the expression of Middle Data $11 = fm11$ (Input Data 111, … Input Data 11i, Middle Data 111, …, Middle Data 11j).

The benefits of the FMEA driven by the data consist of making use the definitions from the system operation concept, and all the FMEA activities apply only on the data, and the process of doing the FMEA above is clearly and completely defined and optimized, the result of which will be efficient, accurate, complete and consistent.

Taking the BSD system that is described at the beginning of this section as an example, the system operation concept is developed as below:

Pixel_Mapping = f_mapping (
        Left_Video_LVDS,
        RGB_Pattern, Clock_Signal);
Frame_Input_Array = f_frame (
        Pixel_Mapping,
        Frame_Input_Time_Constant);
Object_Detected (Object_ID, Object_Label) = f_classification (
        Pixel_Mapping,
        Object_Classification_Array,
        Frame_Input_Array,
        Object_Detect_Time_Constant);
Object_Detected (Object_Relative_Position, Object_Relative_Velocity) = f_detection (
        Pixel_Mapping,
        Object_Classification_Array,
        Frame_Input_Array,
        Object_Detect_Time_Constant,
        CAN_Msg_Veh_Speed);
Object_Detection_Confidence = f_ratio (
        Object_ID,

$$\text{Object\_Label,}$$
$$\text{Object\_Tracking\_State,}$$
$$\text{Object\_Action\_State);}$$

Object_Tracking_State = f_tracking (
        Frame_Input_Array,
        Object_ID,
        Object_Label,
        Object_Traking_Time);

TTC = f_time (
        Object_Relative_Position,
        Object_Relative_Velocity,
        Object_Detection_Confidence);

CAN_BSD_Left_Alert (TTC) = f_output (
        TTC,
        CAN_Msg_Ignition,
        CAN_Msg_Veh_Speed,
        ADC_Veh_Power,
        ADC_Env_Temperature);

The relationships above consist of the following elements:

- Output signals:
  - TTC
  - CAN_BSD_Left_Alert (TTC)
- Input signals:
  - Left_Video_LVDS
  - CAN_Msg_Ignition
  - CAN_Msg_Veh_Speed
  - ADC_Veh_Power
  - ADC_Env_Temperature
- Middle Result:
  - Pixel_Mapping
  - RGB_Pattern
  - Clock_Signal
  - Object_Relative_Position
  - Object_Relative_Velocity
  - Object_Detected
  - Object_Classification_Array
  - Frame_Input_Array
  - Frame_Input_Time_Constant
  - Object_Detect_Time_Constant
  - Object_ID
  - Object_Label
  - Object_Detection_Confidence
  - Object_Tracking_State
  - Object_Action_State
  - Object_Traking_Time

Because all those data are already designed as the key elements in the system, so, those data are the mandatory to do the FMEA; On the other hand, those data fully represent the system, so the FMEA applied to them is complete to reveal the system failure modes. And the relationships between those data are already presented by the system operation concept or the system architecture like the formulas above, i.e., the relationships and the data fully reveal the relationships of failure mode, failure cause and the effect in the system. So, performing the FMEA based on those relationships and data not only fully covers the system failure modes, but also accurately and efficiently collaborates the system development activities.

The next step is to identify the potential failure modes or errors for each data. The strategies of handling data errors are different depending on if the data are Input Signal, Output Signal or Middle results like the list above:

- Input Signal: which are the data that are input from outside the processor using the serial communication protocol which includes the timing parameter:
    - o Data Value Error detection: Since the input signals are from outside, so only the data transmission errors need to be detected, which can be done by checking the CRC or checksum values that are embedded in the signals.
    - o Data Tim Error detection: the input serial signals' timing is defined by the data transmission protocol that is designed in the input signal database, such as the DBC file or XML file, so the input signal arriving time at the input port will be check by following the database parameters.

    The time error cannot be detected for the data input using the parallel communication protocol.
- Output Signals and Middle Result:
    - o Data Time Error: this type error cannot be detected by the developer, because all the data and the instructions that operate the data in a processor will be processed by the ALU in serial, so they don't have the common time as a reference to detect the time error. However, some measures can be done to enhance the code executions that operate the data to ensure the data timing, such as:
        - Internal or external watchdog
        - Window watchdog
        - Carefully design the schedulers and task arrangements
    - o Data Value Error:
        - The data transmission error is detected and corrected by the microcontroller's built-in mechanism, such as Error Correcting Code (ECC), so in most cases, the application does not need to handle this type error, however, if the development needs to run the safety software on the non-safety devices, such as the APU, then the application software should ensure that the non-safety devices meet the safety requirements by explicitly running the data transmission error detection software routines.
        - Data Transforming Error: this type of errors cannot be detected directly because the data are supposed to be changed in a processor, however, some checking can be done to increase the correctness:
            - Range check.
            - Plausibility checks, such as using the experience model,

comparing the data from the simulation.
- Comparing the data using the redundant data storage or redundant calculation results.

In a microcontroller, the binary error described in the section of 2.2.3.1 Reliability may occur to both instruction and data. If the error occurs to the instruction, then the error effect is still some kind of data errors, for example, if the faulted instruction is algorithm operation, then the result data will be wrong; if the faulted instruction is memory operation, then the result location data will be wrong. So, for both cases, only the data error will be handled.

Based on the analysis above, the potential failure modes for the BSD system will be identified as below:

For the Input signals:
CAN_Msg_Ignition and CAN_Msg_Veh_Speed: Those two are the CAN signals, so the failure modes of value errors and the time errors are known from the analysis above, which can be detected by using the CRC values in the CAN protocol and the transmission schedule information embedded in the signals, and the following safety measures will be added to increase the safety:
- adding the signal checksum to against value error
- adding the message counter to protect the message repetition error

So, the actions against those failure modes are:
- The CAN signals' values need to be checked against their checksum values
- The CAN Signals' timing needs to be checked against their transmission schedule
- The CAN message counter value need to match the receiver's counter value.

For example, the two CAN signals from BSD input can be checked as below:
CAN_Veh_Ignition_Status = f_ignstatus (CAN_Msg_Ignition,
CAN_Msg_Ignition_Safety);
CAN_Msg_Ignition_Safety = f_ignsafety (
    CAN_Msg_Ignition_CHECKSUM,
    CAN_Msg_Ignition_Validity);
Among above, CAN_Msg_Ignition_CHECKSUM is to check the signal value error, and the CAN_Msg_Ignition_Validity is to check the signal validity including the timing error.

CAN_Veh_Speed = f_speed (CAN_Msg_Veh_Speed, CAN_Msg_Veh_Speed_Safety);
CAN_Msg_Veh_Speed_Safety = f_speedsafety (
    CAN_Msg_Veh_Speed_CHECKSUM,
    CAN_Msg_Veh_Speed_Validity);
Among above, CAN_Msg_Veh_Speed_CHECKSUM is to check the signal value error, and the CAN_Msg_Veh_Speed_Validity is to check the signal validity including the timing error.

For the signal of Left_Video_LVDS, the similar strategy as the CAN signals will be implemented according to the LVDS protocol as below:
Left_Video_LVDS = f_lvds (Left_Video_Pixel, Left_Video_Pixel_Safety);
Left_Video_Pixel_Safety = f_lvdssafety (
    Left_Video_Pixel_Integrity,

Left_Video_Pixel __Validity);
Among above, Left_Video_Pixel _Integrity is to check the signal value error, and the Left_Video_Pixel __Validity is to check the signal validity including the timing error.

For signal of ADC_Veh_Power and ADC_Env_Temperature, both of them are the analog signals that are measured using the microcontroller's built-in ADC device. Since the analog signal is the continuous signal, so there is no transmission time failure mode.
So, only the data value errors need to be detected, which can only be caused by the ADC device, for which, two type errors need to be detected:
- Device usage error
- Device status error

So, the detection actions will be:
- The ADC configuration shall be checked at the initialization phase.
- The ADC status shall be checked during the measurement.

The ADC measurement deviations can be checked using the internal reference signal provided by the microcontroller:
- The ADC measurement to the reference signal should be check cyclically against the internal reference from the microcontroller.

In addition to the device errors above, for the analog signals, the value out of range shall be detected:
- The measured vehicle power value or temperature value shall be checked against the designed limited range.

Veh_Power = f_power (
   ADC_Veh_Power,
   Veh_Power_Filter_Constant,
   ADC_Veh_Power_Validity);

Among above, Veh_Power_Filter_Constant is ADC filter to reduce the signal value glitch, and the ADC_Veh_Power_Validity is to check the signal validity including the timing error, range error and device error.

Env_Temperature = f_temp (
   ADC_ Env_Temperature,
   Env_Temperature_Filter_Constant,
   ADC_Env_Temperature_Validity);

Among above, Env_Temperature_Filter_Constant is ADC filter to reduce the signal value glitch, and the ADC_Env_Temperature_Validity is to check the signal validity including the timing error, range error and device error.

For the Output Signals and the Middle Result:
For all those data, only the data transforming errors need to be detected, however, this type error cannot be detected by the application software, because the data transformation or data calculation is done by operating some other data to derive the result data, and expected result from those operations is the result data that is under discussion, i.e. the expected results are unknown to the developers, which leads that the pre-defined check action to against the operation results are unknown, i.e. the application developers cannot design specific measures to detect such errors.

Although the fully detections are not feasible, there are still somethings that can be done to reduce the errors:

- Duplication and Comparison

For almost all safety application software, to ensure the correct data operation results, two computer cores are used to execute the same operations for the same data, and compare the results from every step, if there is a difference in any of them, then the error will be reported. In this way, although not all the operation error can be detected, the error opportunities will be reduced significantly.

For example, in 3.4.6.3 BSD Safety, the BSD system uses the RPU0 and RPU1 to run the ASIL B software components, and those two cores are running in the "Lock-Step" mode. In the BSD example, the functionalities allocated on those two cores are:

- o ADC driver in I/O driver of Autosar
- o CAN driver
- o CAN input and output signals: CAN_Msg_Ignition, CAN_Msg_Veh_Speed, CAN_BSD_Left_Alert (TTC)
- o Application Mode Manager
- o Serial Signal Manager
- o Cybersecurity Function
- o Autosar Full contents including the OS, EcuM and BswM components.

This type of Duplication and Comparison is done by the microcontroller's built-in mechanism, for which all automotive microcontrollers have the RPUs consisting of two exact same cores operated by the dedicated management system which manages to run the same instructions and the same date on both two cores, then compare every step of the operations. The core management system will report the error if there is any difference from any operation step, which will result in an interrupt or automatic reset based on the configuration.

If the development needs to run the safety software on the non-safety cores, such as APU or GPU, then the similar solution can be used, which is to execute the same software and same data on two APUs or GPUs, then compare the result data to find out the potential errors.

Another example of such approach is that the vehicle BSD system uses two object detection system: one is the camera BSD, another is the radar BSD, so the vehicle system can compare the detection results from them to make decisions.

- Data Range Check

This method is to check the result data and all or some of the operated data against the designed limitations or rational values, if the data are out of range or unreasonable, then the error will be report.

For example, during the following data calculations:

Pixel_Mapping = f_mapping (
           Left_Video_LVDS,
           RGB_Pattern, Clock_Signal);
Frame_Input_Array = f_frame (
           Pixel_Mapping,
           Frame_Input_Time_Constant);
Object_Detected (Object_ID, Object_Label) = f_classification (

Pixel_Mapping,
Object_Classification_Array,
Frame_Input_Array,
Object_Detect_Time_Constant);

The result data: Object_ID and Object_Label depend on the parameters: Pixel_Mapping, Object_Classification_Array, Frame_Input_Array and Object_Detect_Time_Constant, so the result data: Object_ID, Object_Label, and the middle data: Pixel_Mapping and Frame_Input_Array can be checked about if they are in the reasonable range, which is to reduce the data calculation and transmission errors; and the RGB_Pattern, Clock_Signal, Frame_Input_Time_Constant, Object_Detect_Time_Constant, Object_Classification_Array can be checked against the designed limitations because they are design data, which is to reduce the data transmission error.

For the risk analysis, the classifications of Severity, Probability and Controllability will be used to rate the failure modes and the protection actions which are same as the ones to rate the safety:

Severity (S: 1~4) classifications are defined according to the severity impact of failure modes to the designed functionality:
S0: No Impact
S1: Light to moderate impact
S2: Severe impact
S3: No functionality (the functionality is totally lost)

Probability (E: 1~5) classifications are defined according to the relative expected frequency that the impact can possibly happen:
E0 Incredibly unlikely
E1 Very low probability
E2 Low probability
E3 Medium probability
E4 High probability (impact could happen under most operating conditions)

Severity Class	Probability Class	Controllability Class		
		C1	C2	C3
S1	E1	No Risk	No Risk	No Risk
	E2	No Risk	No Risk	No Risk
	E3	No Risk	No Risk	Low Risk
	E4	No Risk	Low Risk	Medium Risk
S2	E1	No Risk	No Risk	No Risk
	E2	No Risk	No Risk	Low Risk
	E3	No Risk	Low Risk	Medium Risk
	E4	Low Risk	Medium Risk	High Risk
S3	E1	No Risk	No Risk	Low Risk
	E2	No Risk	Low Risk	Medium Risk
	E3	Low Risk	Medium Risk	High Risk
	E4	Medium Risk	High Risk	Very High Risk

Table 3.4-11 Risk Determination

3.4.5 FMEA

Controllability (C: 1~4) classifications are defined according to the relative likelihood that the failure mode can be controlled:
C0 Controllable in general
C1 Simply controllable
C2 Normally controllable
C3 Difficult to control or uncontrollable

Based on the classifications above, the failure modes and their protection actions' risks can be determinate as in Table 3.4-11 Risk Determination, in which the highest risk is rated as "Very High Risk" located at the right-bottom corner, the lowest risk is rated as "No Risk".

In this approach, the classifications of severity, probability and controllability, and the risk ratings determinations are exactly same as in ISO 26262, so that:

- The work products can be re-used: if either the FMEA or Safety ratings following ISO 26262 are established, then they can be used for another.
- The work knowledges can be re-used: if either the FMEA or Safety ratings following ISO 26262 are known, then the knowledge can be used for another.

So, the Data Driven FMEA has significantly simplified the conventional FMEA:

- The selection of the analyzed elements and the elements' structure re-use the work products that are already existed in the system design.
- The risk analysis re-uses the same process that is done for the safety functions.

All data in the BSD example are listed in Table 3.4-12 BSD DD FMEA Analysis, in which, the RPN is based on the Severity (S), Probability (E) and Controllability (C).

#	RPN	S	E	C	Effects (Value Error & Timing Error)	Cause	Preventive action	Detection action
Output Signal								
10	48	4	4	3	TTC	Operation Error Parameter Error	Reliability Design Availability Design	Value Range Check Reasonable Value Check
11	48	4	4	3	CAN_BSD_Left _Alert (TTC)	Operation Error Parameter Error	Reliability Design Availability Design	Value Range Check Reasonable Value Check

	**Value Range Check** **Reasonable Value Check**
21	24	3	4	2	Left_Video_LVDS	Input Device Error Transmission Error	Check Device and Input Protocol	Integrity Check Validity Check
22	24	2	4	3	CAN_Msg_Ignition	Input Device Error Transmission Error	Check Device and Input Protocol	Integrity Check Validity Check
23	24	2	4	3	CAN_Msg_ Veh_Speed	Input Device Error Transmission Error	Check Device and Input Protocol	Integrity Check Validity Check
24	16	2	4	2	ADC_Veh_Power	Input Device Error Transmission Error	Check Device and Input Protocol	Integrity Check Validity Check
25	16	2	4	2	ADC_Env_ Temperature	Input Device Error Transmission Error	Check Device and Input Protocol	Integrity Check Validity Check
	**Middle Signal**
110	32	4	4	2	Pixel_Mapping	Operation Error Parameter Error	Reliability Design Availability Design	Value Range Check Reasonable Value Check
111	8	4	2	1	RGB_Pattern	Storage Error	Design Based on Simulation	CheckSUM
112	8	4	2	1	Clock_Signal	Storage Error	Design Based on Simulation	Clock Reference Check
113	48	4	4	3	Object_Relative_ Position	Operation Error Parameter Error	Reliability Design Availability Design	Value Range Check Reasonable Value Check
114	48	4	4	3	Object_Relative _Velocity	Operation Error Parameter Error	Reliability Design Availability Design	Value Range Check Reasonable Value Check
115	48	4	4	3	Object_Detected	Operation Error Parameter Error	Reliability Design Availability Design	Value Range Check Reasonable

								Value Check
116	12	4	3	1	Object_Classification _Array	Storage Error	Design Based on Simulation	CheckSUM
117	48	4	4	3	Frame_Input_Array	Operation Error Parameter Error	Reliability Design Availability Design	Value Range Check Reasonable Value Check
118	8	4	2	1	Frame_Input_Time _Constant	Storage Error	Design Based on Simulation	Timing Reference Check
119	8	4	2	1	Object_Detect_Time _Constant	Storage Error	Design Based on Simulation	Timing Reference Check
120	48	4	4	3	Object_ID	Operation Error Parameter Error	Reliability Design Availability Design	Value Range Check Reasonable Value Check
121	48	4	4	3	Object_Label	Operation Error Parameter Error	Reliability Design Availability Design	Value Range Check Reasonable Value Check
122	48	4	4	3	Object_Detection_ Confidence	Operation Error Parameter Error	Reliability Design Availability Design	Value Range Check Reasonable Value Check
123	48	4	4	3	Object_Tracking_ State	Operation Error Parameter Error	Reliability Design Availability Design	Value Range Check Reasonable Value Check
124	48	4	4	3	Object_Action_State	Operation Error Parameter Error	Reliability Design Availability Design	Value Range Check Reasonable Value Check
125	12	4	3	1	Object_Traking_Time	Storage Error	Design Based on Simulation	Timing Reference Check

Table 3.4-12 BSD DD FMEA Analysis

3.4.5 FMEA

Since the analyzed elements are related to each other based on the deriving relationship or the formulas, such as: Pixel_Mapping = f_mapping(Left_Video_LVDS, RGB_Pattern, Clock_Signal), the failure modes from the parameters in the formula above have the certain effect to the derived data: Pixel Mapping, and the vice versa is true, as well, i.e., if some failure modes happen to the derived data: Pixel_Mapping, then those failure modes will be related to the parameters if the failure modes are not the intrinsic ones. In another words, the DD FMEA have both the bottom-up logic and the top-down logics.

So, the Data Driven FMEA (DD FMEA) can be used to determine the effect based on the operated data failure modes, and can be used to determine the causes that derive certain consequences, as well, which makes the DD FMEA more useful by comparing the limitation of conventional FMEA that the conventional FMEA is an "Inductive", which limits the conventional FMEA can only be used for bottom-up logic.

3.4.5 FMEA

(Conventional FMEA check list consisting of 8 steps is already described in the Conventional FMEA section)

- Is the system operation concept available that should have all the derivations of system output signals?
- Is the system architecture design available that implements all the derivations of system operation concept?
- Do the error detections cover the input data, middle data and output data about value error and timing error?
- Are the protections implemented for the input data about value error and timing error?
- How are the middle data value errors detected (range check, reasonable value comparison, comparison between duplication) and are the protections implemented?
- How are the middle data timing error protected (watchdog, window watchdog, external watchdog, scheduler design, execution path check) and are the protections implemented?
- What are the output signal error protections implemented (integrity, time)?
- Is the risk analysis done based on the RPN?
- Is the risk prevention strategy (such as design quality control) defined based on the RPN?
- Is the risk improvement strategy defined based on the RPN changes?

3.4.6 Safety

The automotive vehicle safety involves two development aspects:
- Safety of the intended functions (SOTIF) guided by ISO 21448
- Conventional system safety guided by ISO 26262

The safety of the intended functions is to ensure that the intended functions, such as the Lane Keeping (LK) functions, Assist Emergency Braking (AEB) functions, Blind Spot Detection (BSD) functions, are used properly.

All those functions need the help of environment sensors, such as radar sensor, camera sensor, to detect the vehicle surrounding situations during driving, such as the objects on the path (vehicles, pedestrians, bicycles) and the road on the path (surface, lane, guardrail, overhead) to execute the intended functions properly. And those sensors have their performance limitations, such as how far and in what horizontal and vertical angles the radar sensors or camera sensors can detect objects correctly, and the detections will be impacted as well by some environment factors, such as rain, snow, time of day.

And some system performance will depend on the implementations, for example, for the Lane Keeping functions, some systems will totally depend on if the sensors can detect the lane, which cannot work correctly if the lane marks are not clear or the marks are covered by something; while other system will predict the lane not only using the sensors but also using map or other means to increase the system performance.

So, to be safe, the intended functions must be used within their performance limitations, otherwise it may cause potential safety issues.

The conventional system safety development is to prevent the systems from internal faults, such as random hardware faults, implementation mistakes. For example, the braking system in a vehicle needs to prevent from the internal faults, such as the memory devices issues that may cause the unintended braking during driving at the certain speed, which may cause potential hazards; the steering system in a vehicle needs to prevent from the internal faults, such as the microcontroller execution issues that may cause the unintended lateral moving, which may cause potential collisions.

3.4.6.1 Safety Concept

What are the vehicle safe driving situations?
What are the safety requirements in an automotive ECU system?
Where and how should the safety requirements be derived from?

By following ISO 21448, both following vehicle operation scenarios must be handled:
- Ego vehicle operates outside the system performance limitations
- Ego vehicle operates within the system performance limitations

So that, the safety of automotive ECU systems especially the automated driving systems can be extended to cover more situations where not only the systems' internal faults but also the vehicles' misuses without any internal fault should be prevented. Which can be covered by:
- To identify the vehicle performance limitations and to avoid misuses by validating the systems' sensing and acting abilities against the road users and roads considering environment impacts such as geography, time of day, weather, infrastructure including the traffic signs.

- To develop the indicators (visual, auditory and haptic display) and the user instructions to warn the misuses once the ego vehicle operates outside the system performance limitations
- To development the internal safety mechanisms to prevent the systems from internal failures when the ego vehicle operates within the system performance limitations by following ISO 26262.

The prerequisites of the internal safety mechanisms developments are the safety requirements developments, and all parts of which in a vehicle are derived from the vehicle safety requirements, which in turn are from the law:

"(8) "motor vehicle safety" means the performance of a motor vehicle or motor vehicle equipment in a way that protects the public against unreasonable risk of accidents occurring because of the design, construction, or performance of a motor vehicle, and against unreasonable risk of death or injury in an accident, and includes nonoperational safety of a motor vehicle (From: MOTOR VEHICLE SAFETY, TITLE 49, UNITED STATES CODE, CHAPTER 301)".

The legal safety requirement above can be divided into the following:

- Safety Requirement 1: the vehicle shall ensure the people safety who are on the vehicle' path.
- Safety Requirement 2: the vehicle shall ensure the people safety who are inside of the vehicle

Those two safety requirements are cascaded to every part in a vehicle, including ECUs.

Based on above and the Data Driven concept, this book provides the safety development approach in an electronic system, especially in the automotive ECU systems in a more complete and comprehensive manner, which is partially compliant to ISO 26262 with some differences.

One of differences is the safety requirement derivation. Although the safety requirements are derived from the higher-level components in ISO 26262 as well, the derivation is based on the hazard analysis, which is the "indirect" way to derive the required safety requirements.

For example, at the beginning of Clause 6 in the Part 3 of ISO 26262:2018, it states: "The objectives of this clause are: a) to identify and to classify the hazardous events caused by malfunctioning behavior of the item; and b) to formulate the safety goals with their corresponding ASILs related to the prevention or mitigation". So, the development sequence in ISO 26262 is to do the hazard analysis and risk assessment first, then derive the safety goals and safety requirements.

However, in this book, the items' safety requirements should be directly derived from the higher-level items' safety requirements or directly from the vehicle safety requirements. And the items' partial safety requirements are to prevent from internal failures that are same as the ones required by ISO 26262, the other parts of safety requirements are the ones to execute the intended safety functions. So, all safety requirements of items are to serve the intended safety functions (braking, steering), rather just prevent from internal faults.

In such way, the development work flow will be clearer and more straightforward as required by ASPICE.

The requirements for an electronic system include the following three aspects:
Functionality

Reliability
Quality

Both ISO 26262 and IEC 61508 address only the reliability and the reliability related quality.

The relationships between the three aspects are:

- The functionality is what is required, which is represented by the functional availability.
- The Reliability is the measure to ensure the correctness of required functionality, which focuses on that the system will not function wrongly if something is wrong in the system.
- The quality is the management measure to reduce mistakes in the system development.

If an electronic system is the safety relevant system, then all three aspects above are safety topics, because both functionality and reliability are the direct aspects related to the safety relevant functionality delivery, and the quality control is a safety relevant development control, as well, all the activities in those three aspects serve the system final behaviors.

Both ISO 26262 and IEC 61508 are the development standards or guidelines based on the "hazard analysis", which the system needs to develop the mechanisms to prevent the system from those analyzed hazards to ensure the correctness of system functionality, such mechanisms are well known as the "fail-safe" measures, for the development of which, those two standards are quite comprehensive. However, they do not cover all the safety topics in a safety system development. The potential issues of which are:

- It may miss some safety requirements that are directly inherited from the higher-level components.
- It may degrade the system functional availability.
- It may confuse the safety analysis, especially the safety requirement derivations:
    - The main safety requirements in an electronic system, especially in the automotive ECU systems, are inherited from the higher-level systems, which are finally rooted from the vehicle.
    - The safety requirements that are derived based on the hazard analysis by following ISO 26262 or IEC 61508 are the ones that are, in fact, indirectly derived from the safety requirements above, as well. So, the safety analysis will be more straightforward if the safety requirements are derived from the higher level.

For the example of "It may miss some safety requirements", for the braking system in an automotive vehicle, the requirement of "the braking system shall output the braking force of F Newtons in T seconds if the brake pedal angle below m C and the angle change ratio is N C / S". It is a main functionality requirement and most importantly it is a safety requirement to the brake system because the brake system is a very important safety system in a vehicle no matter there are hazards or not in the system.

If the brake system could be implemented in the "perfect" platform that the system will always output the required force in all situations, then the requirement above is still a safety requirement to the system because the braking is the safety measure to the vehicle though there is not any hazard in the braking system, and the safety relevant development quality control must still be engaged. So, not all safety requirements are derived from the hazard analysis, in another words, the approach based on the hazard analysis may not cover all safety requirements in all situations.

For the example of "It may degrade the system functional availability", In the safety requirement engineering, If the ASIL decomposition is used and the ASIL decomposition of an initial safety requirement results in the allocation of decomposed requirements to the intended functionality and an associated safety mechanism, then the associated safety mechanism should be assigned the highest decomposed ASIL (Clause 5.4.7 in ISO 26262-9). The approach of which will make sure that the system will not go wrong because the associated safety mechanism will have the qualified resources (platform hardware and quality control), however, the intended functionality may not be assigned to such qualified resources and quality control, so, the system may not be able to deliver the required functionality all the time, i.e., the system functional availability may be degraded, and the functionality is the major part of the system ability to ensure the vehicle safety. So, if the safety development in an electronic system, especially in an automotive ECU system is based only ISO 26262 or IEC 61508 without considering the whole picture, it may degrade the system functional availability.

For the point of "the safety analysis will be more straightforward if the safety requirements are derived from the higher level", the benefits and values of such safety requirement derivation are that the safety development procedures are same as the normal system development:

- o To an electronic system, the requirements including the safety requirement should be given in the way such that the requirements describe the system input and output signals and the functionality from the system external behavior point of view, and the requirements are derived from the higher-level system functionality using the decomposition and allocation in ISO 26262.
- o Inside of the system, the internal design items including the reliability mechanisms are derived using the approach in 3.4.1.1 Derivation:
    - The sub-systems derive their requirements by decomposition.
    - The mechanisms that ensure the correctness of system functionality are the derived requirements according to ISO 26262 or IEC 61508.

Another issue in the safety development is that the criteria to measure if a system is safe are very vague, although there are some quantitative hardware criteria from ISO 26262 about Single Point Fault Metric (SPFM), Latent Fault Metric (LFM) and the Probabilistic Metrics for Hardware Failures (PMHF) listed in the Table 3.4-15 Hardware Fault Metrics:

- The system safety measurements, especially the software safety measurements are not specified.
- Only very high-level activities are specified in ISO 26262 Part 4: Product development at the system level and the ISO 2626 Pat 6: Product development at the software level, but those activities are highly dependable to interpretation and implementation, which is very difficult to make accurate adjudgment.

To improve it, this book provides solutions from the Data Driven point of view:

- Full system error detection: provide the solution to fully detect the errors in the system under development, which consist of only two types of errors: value error and timing error.
- Approach to achieve the safety: provide the solution to achieve the three aspects that are needed by the safety in the system under development.

- ISO 2626 compliance: provide the rationales for the solutions in the book to meet the ISO 26262 requirements, which can be a leading example for similar projects to achieve the compliance.

3.4.6.2 Safety Development

As described above, the automotive vehicle safety involves two development aspects:
- Safety of the intended functions (SOTIF) guided by ISO 21448
- Conventional system safety guided by ISO 26262

Safety of the intended functions (SOTIF)

The key point of SOTIF is to identify the system performance limitations, by which, the vehicle safety development can be classified as below:
- Outside the limitations (unknown, both hazardous and non-hazardous)
  - o Solution: User Manual / Instructions
  - o Misuse indicators (both visual, auditory and haptic display) in the vehicle
- Within the limitations (ISO 26262 scope)
  - • both known and unknown
  - • Both hazardous and non-hazardous

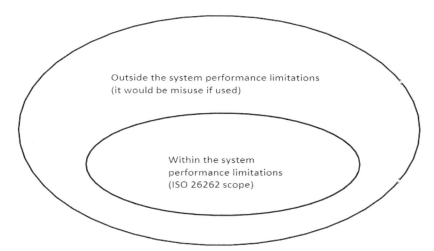

Outside the system performance limitations
(it would be misuse if used)

Within the system
performance limitations
(ISO 26262 scope)

Figure 3.4-27 SOTIF Classification

In the SOTIF development, the following assumptions are applied:
- Safety Definition:
  - o Don't hit any dynamic road user.
  - o Stay on planned track (e.g., the planned vehicle track shall avoid dangerous obstacles)
- To operate vehicle safely:
  - o The ego vehicle shall be able to both act to the road users and be on planned track safely, for which the ego vehicle should be able to sense both the road users and the road within the performance limitations.
  - o The ego vehicle operator(s) need to know the performance limitations.

- "known / unknown" means if the ego vehicle is used within the performance limitations (not the driver who is responsible only for action. If the driver is responsible for the detection, the SOTIF is the conventional vehicle functions like braking / steering; if the vehicle designers don't know the performance limitations, then the results are unknown)

And to identify the systems' performance limitations, the following actors should be considered:

- Subject (ego vehicle, driver, passengers)
- Object (Road users (Vehicle, pedestrian/animal), Road (surface, lane, guardrail, overhead))
- Environment (Geography, time of day, Weather, Infrastructure including the traffic signs)

To identify the systems' performance limitations, the situations in the table below need to considered and to be validated, if all the situations have the "Yes" answers, then system is within the performance limitations, otherwise, it is outside the limitations which should be prevented.

		Object under Environment	
		Dynamic Road User	Road (surface, lane, guardrail, overhead)
Subject	Sensing (by vehicle)	Yes / No	Yes / No
	Plan (by vehicle)	Note: "Plan" will always have the default path (Designed Safe State) if plan is not feasible, so it will always be within the performance limitations.	
	Action (by both vehicle and driver)	Yes / No "Yes" means that the ego vehicle has the ability to act to the road users safely (e.g, braking, steering, external airbag)	Yes / No "Yes" means that the ego vehicle has the ability to be on the planned track (e.g., stay in the lane)

Table 3.4-13 Performance Limitation

Conventional system safety

For the conventional system safety development, this book provides the Data Driven approach. In the Data Driven approach, the system functionalities are presented by the output data illustrated in Figure 3.4-1 Computing System with Multiple Signals, in which the relationships between the output data and the input data can be described by the formulas below:

Output Data 1 = $f1$ (Input Data 11, …, Input Data 1i, Middle Data 11, …, Middle Data 1j);
Output Data 2 = $f2$ (Input Data 21, …, Input Data 2l, Middle Data 21, …, Middle Data 2p);

• • •

Output Data n= fn (Input Data n1, …, Input Data nq, Middle Data 1n, …, Middle Data nr).

So, the safety goals are to ensure the system output data to meet the requirements, which, in turn, requires that the middles result and the input signals must meet the safety requirements, as well.

To realize the safety goals, there are three system safety concepts that should be covered:

- Reliability
- Availability
- Quality

Reliability:

The system Reliability is the system ability to provide the implemented functionalities.

In the section of 2.2.3.1 Reliability, there are the example and the explanations about the system reliability.

To ensure the system reliability, the system needs to implement the mechanisms to detect the potential errors, which is covered by the FMEA, from which, there are two types of errors:

- Human error, which is the mistakes made by the developers, which are mainly handled by the quality control.
- Systematic errors, which are the non-foreseeable and random errors that will be handled by the FMEA, which has only two types of errors: Value error and Timing error.

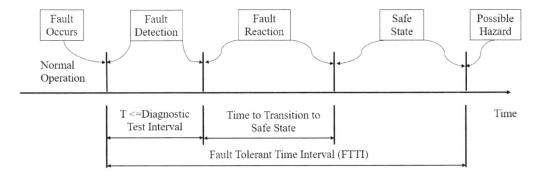

Figure 3.4-28 Fault Tolerant Time Interval (FTTI)

Once the error detection mechanisms are developed, the system needs to take suitable actions to prevent the system from the wrong behavior that is caused by the errors:

- The system needs to go to the designed Safe State in the designed time, in which, the system needs to:
  - stop the functional activities, i.e., stop sending the output data
  - send out the warning to indicate the error state
  - store the errors information for the diagnosis later
  - take actions to recover

3.4.6 Safety

The definition of the system reliability from the data driven point of view: the system reliability is that the system must not send out the wrong data, i.e., in the cases where the system cannot send out the correct data, it must stop sending out the data and provide the warning.

The time for the system to go to the Safe State after detecting errors must be within the Fault Tolerant Tim Interval (FTTI) illustrated in Figure 3.4-28 Fault Tolerant Time Interval (FTTI) above, in which:

- There is the time gap between the time when the fault occurs and the time when the fault is detected, which is defined as the Fault Detection or Diagnostic Test Interval.
- After the time of Fault Reaction Time, the fault detection can find and confirm the fault, the execution will go to the safe state.
- In the safe state, the first reaction is to stop sending out the functional signals and sending out the warning signals if the communication interfaces still work.
- After the receiver receives the warning signals or detect that the required functional signals are lost, the receiver can react to such situation.

The FTTI is a critical requirement to the system safety and should be designed carefully, the design principle is the shorter, the better.

All errors must be recorded by the system for the diagnostic reason, which will be done using the AUTODAR DEM functionality, and all that information must be reported to the users for the repair and product analysis reasons, which will be done using the Diagnostic Trouble Code (DTC), which is the project specific and will be implemented by the application.

The system actions to prevent the system from the errors will depend on the errors categories, and there are following categories of errors and handling approaches in the automotive systems:

- Hardware errors that are detected by the microcontroller itself

This type of errors is detected and reacted by the microcontroller built-in functions; the reaction is to reset the microcontroller.

During the power on reset and normal execution, an automotive microcontroller will run the built-in self-tests to check and make sure that all devices in the microcontroller are in the normal status. Once the errors are detected, the built-in functions will store the errors' information in the dedicated location, then reset, after which, the application can retrieve the errors' information to diagnosis.

- Hardware errors that are detected by the applications

This type of errors is mainly from the external devices connected to the microcontroller, such as the communication interface components (CAN transceiver, Ethernet transceiver, Seriralizer, Deserializer, etc.), circuits of AD / DA / DIO, which are illustrated in Figure 2.2-2 ECU Input and Output Signals.

Those devices and circuits don't belong to the microcontroller, so the microcontroller does not have knowledge to test them, so the application software must take care of them. The FTTI for this type of errors is usually 130 ms ~ 180 ms, and the reaction to recover from the errors is to go to the safe state, then reset.

- Software Errors that impact the system safety and cannot be recovered by the application.

This type of errors is mainly the errors discussed in the section of 3.4.5 FMEA consisting of the Data Value Error and the Data Time Error.

If the errors occur in the data related to the safety output data, then they impact the system safety, for such errors, the reaction is to go to the safe state as soon as possible to stop the output data.

The FTTI will depend on the data type and the error type:

- Serial communication input data value error: since the data transmission rate is project specific, though the error action time can be designed, so the FTTI will depend on the input signal transmission rate.
- Serial communication input date time error: this type of errors, such as the input signal is lost, will depend on the input data transmission rate, as well, usually the data will be defined as lost if the data does not occur in 3.5 periodic times. And each serial communication input signal has different periodic time, so the time to detect the signal lost is different, as well.
- Locate data time errors: this type of errors is detected by the watchdogs, since different software task has different watchdog timeout setting, the FTTI will be different.
- Locate data value errors: if this type of errors is detected using the mechanism mentioned in the section of 3.4.5 FMEA, for example, then the FTTI should be from 180 ms to 300 ms depending on the system requirements.

- Software Errors that impact the system safety and can be recovered by the application.

If this type of errors can be recovered immediately, then the system does not need to go to the safe state, rather, recover the error. For example, there are two System Power Mode signals in a vehicle, if the main System Power Mode signal is wrong, then backup System Power Mode signal should be used.

If this type of errors recovery takes time that may lead to that the system may output wrong data, then the system must go to safe state first, then take the recovery action.

- Software Errors that don't impact the system safety

This type of errors does not need to go to the safe state, the reactions are project specific.

- Environment conditions are out working range

This type of errors includes the environment temperature and the vehicle input power supply voltage out of range, which are detected using the microcontroller's built-in ADC devices plus the analog signal conditioning circuits. Once the errors are detected, the system needs to go to the safe sate because the whole ECU cannot work normally under such out of range conditions, however, there is not any recovery action to be taken by the microcontroller because this type of errors cannot be recovered by the microcontroller itself. What the microcontroller needs to do is to stay in the safe state and monitor the environment conditions to see if the conditions are back to normal range.

Once the execution is in the safe state, the last action in it is to recover from the fault, which can be done by:

- Task Rollback

This type of recovery has the minimum impact to the system, which is to take actions based on the return status from the task consisting of:

- o Taking the design default values, or the backup values. For example, the NVM manager will take the backup values if the reading values are wrong.
- o Re-execution the task again until the failure reaches the design threshold value. For example, the NVM manager will try 3 times if the writing to the NVM unsuccessfully.

- Memory Partition Reset

This type of recovery will handle the errors, such as the memory access violation, execution watchdog timeout, etc., and impact only the components in the partition. As described in the section related to the Freedom From Interference (FFI), the BswM will collaborate with the EcuM to shut down and restart the partition.

- Processor Core Reset

This type of recovery will handle the errors, such as the process shared memory, peripherals and the core specific devices' controllers, and will impact all the partitions in the core. The EcuM will collaborate with the Master EcuM to shut down and restart the specific core. All the BswM in the core need to prepare and clean up for the core reset.

- Microcontroller Reset

This type of recovery will handle the errors that impact the whole ECU, such as the external ADC circuit errors, the microcontroller startup built-in check errors. The master EucM will collaborate with the bootloader to reset the ECU.

Availability

The availability is the ability to provide the required functionalities even something goes wrong. To achieve the goal, there are two ways to design the system:

- Duplication: it requires that the product should have the redundant mechanisms for certain important functionalities, and those redundant mechanisms must be independent each other, so that in the cases where one of them is in the failure mode, then another or others can still provide the required functionalities.
- Recovery from Faults: it requires that the product should recover from the faults in the manner that the system functionality is still acceptable while the safety is not impacted.

For example, in most autonomous driving vehicle systems, there are both camera and radar ECUs to detect the front objects, each of them has totally different mechanism for the object detection and has its own communication channels to the host vehicle.

For the example BSD system, the FPGA will implement the backup output warning signal mechanism which duplicates the one from the APU; The external NVM data will be duplicated by the AUTOSAR NVM data block backup functions; The CAN transceiver will be reset if the CAN bus off error is detected.

The system's availability, costs and latency performance conflict each other, i.e., increasing the availability will increase the system cost and the system latency, as well. So, to achieve the common goals is one of the systems engineering critical points.

Quality

The goals of quality control in the development are to make sure that the development:

- Designed what are required
- Implemented what are designed

In another words, quality control is to prevent the development from the mistakes, which can be done from two aspects:

- From the technical aspect
- From the management aspect

From the technical aspect, the method to ensure the development quality is to do the Verification or Test, and the method to test a product is:

- To run the product if it is a piece of software source code
- To simulate the product if it is a design concept
- To review the product if it is a document
- To plan the activities and review the execution if it is a development procedure

From the management aspect, the quality assurance is to have the qualified development process in place, which demonstrates that the ECU developers have the "good faith" and the organizations have the established procedures to develop the safety products, which should include:

- The established development processes, such as ASPICE, to direct the new product developments, or re-use the existed product and technology for the development.
- The established development processes are compliant with the industrial standards, such as ISO 26262, UN ECE 155 / 156.

3.4.6.3   BSD Safety

This section will take the left BSD system example which is described in the section of 3.1.3 Example as the safety use case to demonstrate the Data Driven Safety development.

Although the use case is close to the real application, it must not be used as in the real application because there are quite some measures that are not designed in it. Furthermore, there are two significant safety issues that exist in both this example and the most real applications:

- This left camera BSD sub-system has the ASIL B(D) safety rating that is decomposed from the vehicle safety requirement for the BSD to prevent the vehicle from the left turn collision that is an ASIL D. However, according to ISO 26262, the ASIL B(D) system hardware must be compliant with ASIL D, which most camera object detection system cannot meet.
- Even regardless the hardware ASIL D requirement, the left camera BSD system safety requirement ASIL B cannot be me, either, because this left camera BSD system and most object detection system in the market currently have only the 95% or 97% confidence rating in maximum, i.e., the systems can only detect the 95 or 97 objects out of 100 objects, which is 10 million times less than the ISO 26262 ASIL B requirements. The analysis is in the section of "Safety issue in autonomous driving based on ISO 26262" of 3.4.6.4 ISO 26262 Compliance.

The two safety issues above need to be carefully considered in the real applications, which is one of the key points from this book.

In this user case, the safety requirement analysis approach is Top-Down:

- First Step: the performance classification starts from the vehicle level

- Next Step: the safety requirement analysis starts from the vehicle level
- Next Step: the assembly level safety requirements that are derived from the vehicle
- Next Step: the ECU level safety requirements that are derived from the assembly
- Next Step: the processor level safety requirements that are derived from the ECU
- Next Step: the partition level safety requirements that are derived from the processor
- Next Step: the function level safety requirements that are derived from the partition
- Next Step: the data level safety requirements that are derived from the function

Performance Classification:

The left BSD radar sensor and camera sensor performance limitations should be classified about how far and in what horizontal and vertical angles the radar sensor and the camera sensor can detect objects correctly, and the detections will be impacted as well by some environment factors, such as rain, snow, time of day.

Safety at the vehicle level:

The BSD system safety ASIL level is derived from the vehicle safety goal about the Blind Spot Detection.

Hazard and safety goals related to Left BSD:

Hazard: Potential collision between the host vehicle and the left-behand purchasing vehicle.

Safety Goal at vehicle level: the left BSD shall prevent the host vehicle to make left turn if there is the potential collision and provide the warning to the drive.

Safety Requirements at the vehicle level:

- The left BSD system shall prevent the host vehicle to make the left turn if the Time To Collision (TTC) between the host vehicle and the detected vehicle in the left BSD detection zone is less than 2.5 seconds (ASIL D).
- The left BSD system shall warn the driver if the Time To Collision (TTC) between the host vehicle and the detected vehicle in the left BSD detection zone is less than 2.5 seconds. (ASIL C).

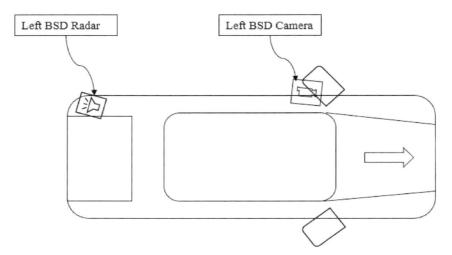

Figure 3.4-29 Left BSD Decomposition

Safety at the assembly level:

The Blind Spot Detection assembly consists of:

- Object detection subsystem
- Power steering control subsystem

The assembly of left BSD is an ASIL D system, the Electronic Power Steering Module (EPSM) was developed as the ASIL D in the vehicle, and the object detection must be the ASIL D, as well.

However, all of camera and radar object detection systems in the current market are developed as ASIL B, so to nearly satisfy the safety requirement, the left BSD object detection requirement is decomposed to two sub-system: camera and radar system, each of which is ASIL B (D).

So, the vehicle left BSD system consists:

- Vehicle Dynamic Control
    - Electronic Power Steering Module (EPSM), which is ASIL D system.
- Driver Warning
    - HMI for the warning signals (sound and light), which is ASIL C system.
- Left BSD Object Detection
    - Camera System
        - Left BSD camera, which is ASIL B (D)
        - Left BSD System ECU, which is ASIL B (D)
    - Radar System
        - Radar Sensor with Object List output, which is ASIL B(D)

The safety issue is that neither the camera nor the radar hardware in the current market meets the ASIL D requirements. According to the clause 5.4.5 in Part 9 of ISO 26262:2018, the requirements specific to the random hardware failures, including the evaluation of the hardware architectural metrics and the evaluation of safety goal violations due to random hardware failures remain unchanged by ASIL decomposition.

And this type of issues currently exists in most of autonomous driving environment sensing devices in the real applications.

Safety at the ECU Level:

Left Camera BSD System ECU is rated as ASIL B (D).

Requirement analysis for the left camera BSD system ECU (camera object detection):

(In the following, for the simplicity, the left camera BSD system is referred as either the left BSD or BSD system.)

The entire left BSD system TTC is 2.5 seconds, and the entire system latency including both the Object Detection system and the Power Steering system is 0.524 seconds according to Table 3.4-9 BSD Signal Latency, so the entire left BSD system must detect the target vehicle at least at 3.024 seconds before the potential collision.

If the speed deviation between the subject vehicle and the target vehicle is 100 KM/S, for example where the subject vehicle is merging onto the highway, then the distance to detect the target vehicle is about 84 meters

For the left BSD ECU system, the latency from the input signal arrived at the input port to the output signal out of the output port is total of 232 milliseconds.

Hazard 1: the left BSD ECU system does not meet the requirement that the system shall detect the target vehicle in the detect zone and send out the required signals in the required time (232 ms).

Hazard 2: the left BSD ECU system sends out unintended object detection signals.

Safety goals:
Goal 1: Under the enabling conditions, the left BSD ECU system shall detect the target vehicle in the detect zone using the camera input pixel signals and send out the required signals (CAN_BSD_Left_Alert (TTC)) in the 232 ms that is measured from the first camera input pixel arrived at the input port to the first output signal out of the BSD ECU, which is ASIL B (D).

Goal 2: the left BSD ECU system shall not send out the unintended object detection signals (CAN_BSD_Left_Alert (TTC)), which is ASIL B (D).

The two safety goals are exactly covered by the BSD system's output data attributes: Data Value and Data Timing.

Safety at the data flow level:
In Data Driven System Engineering approach, all functionalities are represented by the data, and the relationships between them are represented by the formulas in section of 3.4.1 System Operation Concept Design, for the BSD system, the data and the relationships are listed below. Based on the analysis at the beginning of this section that the system safety will focus on the data about value error and timing errors, and the safety requirements for each data will be derived by following those data and the relationship.

In the safety requirement engineering based on ISO 26262, if the ASIL decomposition is used and the ASIL decomposition of an initial safety requirement results in the allocation of decomposed requirements to the intended functionality and an associated safety mechanism, then the associated safety mechanism should be assigned the highest decomposed ASIL (Clause 5.4.7 in ISO 26262-9).

For example, in the analysis of Enabling signal train below, there is the dependency of CAN_Veh_Ignition_Status = f_ignstatus (CAN_Msg_Ignition, CAN_Msg_Ignition_Safety), in which, the CAN_Veh_Ignition_Status is one of enabling conditions to the BSD system, so it is ASIL B data, which can be decomposed to

- Retrieve the signal CAN_Veh_Ignition_Status from the CAN message of CAN_Msg_Ignition, which is QM
- Prevent the CAN message of CAN_Msg_Ignition from the value error and timing error, which is ASIL B.

However, such decomposition has the issues:

- It may degrade the function availability of retrieve the signal CAN_Veh_Ignition_Status from the CAN message of CAN_Msg_Ignition if the function is implemented as QM according to the analysis in 3.4.6.1 Safety Concept.
- It does not simplify the development, because if the QM function and the ASIL B function are allocated into different partitions, then such decomposition would make sense. However, those two functions naturally work together and it is easy to development them together, so the separation will increase the development complexity and decrease the system performance by adding the latency.

So, in the following analysis, if the separation does not add development value, then it is not needed.

In the left camera BSD system, there are the following known data flow according to 3.4.1.2 BSD Example:

- Output date flow: CAN_BSD_Left_Alert (TTC);
- Object detection date flow: TTC, Object_Detection_Confidence, Object_Tracking_State, Object_Detected, Object_Detected, Pixel_Mapping, Frame_Input_Array
- Enabling date flow: CAN_Veh_Ignition, CAN_Veh_Speed
- Camera input pixel data flow: Left_Video_LVDS;
- Environment date flow: Veh_Power, Env_Temperature;

All data in the data flows above are rated as ASIL B except the Environment signals: ADC_Veh_Power and ADC_Env_Temperature because those two signals' measurement has very high toleration about measurement time and measurement precision, so, their impact to the output signal is not critical, for example, the temperature signal will not have much different impact to the output signal between the +84°C or +86°C, and the signals have the similar toleration about measure time, as well.

All other data have the impact to the output signal: CAN_BSD_Left_Alert (TTC), because those data derive the output signal either directly or indirectly, and according to the clause 7.4.2.2 of ISO26262-4: "Each element shall inherit the highest ASIL from the technical safety requirements that it implements." So, all those data are rated as ASIL B.

In the Safety analysis above, the ASIL identification or the safety requirement for each data is straight forward based on the implementation relationship to the output signals, which simplifies the safety analysis. And in the Table 3.4-12 BSD DD FMEA Analysis, the safety mechanisms for each data in the BSD system are designed by the preventive actions and the detection actions.

Based on the safety requirements and design for each data, function, partition, processor and for the whole ECU, then the safety mechanism development including the safety mechanisms allocations can be carried out according to the system architecture design, in which the safety and the reliability mechanisms will be allocated onto the individual component.

Safety at the processor level:

All the functionalities are allocated to the dedicated components based on the structure of Figure 3.4-7 ECU Component Structure. The BSD system as the example will use a microcontroller like the one in the section of 3.1.3 Example, in which:

- There are two RPU core running in the "Lock-Step" mode for the safety tasks.
- There are two APU cores for the high-performance tasks.
- One GPU with the Pixel processor
- The FPGA provide the control logic for audio interface, which is the redundant output warning signal in this application.
- The SHE unit provide the crypto algorithm engine for the cybersecurity functions.
- Internal RAM and EEPROM provide the memory for the middle result storage.

The functionality assignments and safety requirements can be allocated as following:

RPU0 and RPU1:
- Two partitions: one for ASIL B and another for QM, the detailed allocations are

described later in the "Safety at the partition level" section below.

GPU:
- Only one partitions, which is ASIL B;
- Pixel_Mapping = f_mapping (Left_Video_LVDS, RGB_Pattern, Clock_Signal);
- Frame_Input_Array = f_frame (Pixel_Mapping, Frame_Input_Time_Constant);
- The Complex Device Drive (CDD), which is the LVDS driver function for the Left_Video_LVDS;
- Autosar: OS, EcuM, BswM;

APU0:
- Only one partitions, which is ASIL B;
- Object_Detected (Object_ID, Object_Label) = f_classification (
  - Pixel_Mapping,
  - Object_Classification_Array,
  - Frame_Input_Array,
  - Object_Detect_Time_Constant);
- Object_Detected (Object_Relative_Position, Object_Relative_Velocity)
  - = f_detection (
  - Pixel_Mapping,
  - Object_Classification_Array,
  - Frame_Input_Array,
  - Object_Detect_Time_Constant,
  - CAN_Msg_Veh_Speed);
- Autosar: OS, EcuM, BswM;

APU1:
- Only one partitions, which is ASIL B;
- TTC = f_time (
  - Object_Relative_Position,
  - Object_Relative_Velocity,
  - Object_Detection_Confidence);
- Object_Detection_Confidence = f_ratio (
  - Object_ID, Object_Label,
  - Object_Tracking_State,
  - Object_Action_State);
- Object_Tracking_State = f_tracking (
  - Frame_Input_Array,
  - Object_ID,
  - Object_Label,
  - Object_Traking_Time);
- Autosar: OS, EcuM, BswM;

FPGA:
- Backup Object Detection Warning Signal: FPGA_BSD_Left_Alert (TTC), ASIL B;

GPU and APU safety enhancement:

Usually, the GPU ad APU are not suitable for the safety applications, but nowadays as the autonomous driving progresses, more and more camera and radar applications use those high-performance processors to process the object detection, some of them need to meet the safety requirements like the BSD example if the application is the safety relevant.

To meet the ASIL B safety requirements, the following safety measures need to be considered:

- Platform Management Unit (PMU): In the automotive microcontroller, there is always the PMU implemented to monitor the processors in it, which includes the platform management, memory management unit (MMU), peripheral management unit (PMU).
- Watchdog to monitor the PMU: This measure prevents the PMU from execution errors.
- Clock monitor: This measure prevents the time errors which cannot be done by the application that is mentioned in the section of 3.4.5 FMEA.
- BIST: all microcontrollers have the Built in Self-Test (BIST) for the RAM, ROM, communication bus control register, logic control register, interrupt control register, etc., those BISTs are either built in the device firmware or in the software library, the application component needs to invoke them during the power on initialization phase and the operation phase to monitor the processors, so that the platforms can meet the safety requirements.

Safety at the partition level:

RPU0 and RPU1 with two partitions: one for ASIL B and another for QM. Those two cores are running in the "Lock-Step" mode, in which, the two cores will execute the same instruction set and data, then compare the execution result to detection any potential errors. Under the ARM architecture, such execution mode can reach the ASIL D requirement about 99% SPFM, which satisfies the safety requirement ASIL B(D) fully.

For the BSD example, the functionalities allocated on those two cores are:

- ADC driver in I/O driver of Autosar
- CAN driver
- CAN input and output signals: CAN_Msg_Ignition, CAN_Msg_Veh_Speed, CAN_BSD_Left_Alert (TTC)
- Application Mode Manager
- Serial Signal Manager
- Cybersecurity Function
- Autosar Full contents including the OS, EcuM and BswM components.

All those functionalities above are rated as ASIL B and executed in the ASIL B partition. The EcuM component here is the master EcuM which is started by the bootloader during the power on reset, and it controls other EcuMs in other cores starting up and shutdown.

Those components are allocated in the QM partition.

- Diagnostic Services
- Environment signals: ADC_Veh_Power and ADC_Env_Temperature
- The I/O HW abstraction and driver are QM

The communications between the QM partition and the ASIL B partition are safe guarded as below:

- In the default diagnostic session: Only the information query requests will be executed, so, the QM query will not interfere the ASIL B contents.
- In the Programming and Extended diagnostic sessions, the vehicle status will be safe guarded to make sure that the vehicle is not moving where the BSD feature function will not function, so the freedom from interference and the safety are ensured.
- The vehicle environment temperature and power supply monitoring is not timing and precious critical, and the output signals from them are continuously to the environment enabler, so they are rated as QM.

All the processors above will use the shared memory to communicate with each other, in such way, the communication latency between each processor will be minimum, so that the system latency will depend on each processor' main task scheduling time.

The system latency is described in Figure 3.4-22 BSD System Latency and Table 3.4-9 BSD Signal Latency of section of 3.4.4.3 Latency.

Safety at the function level:

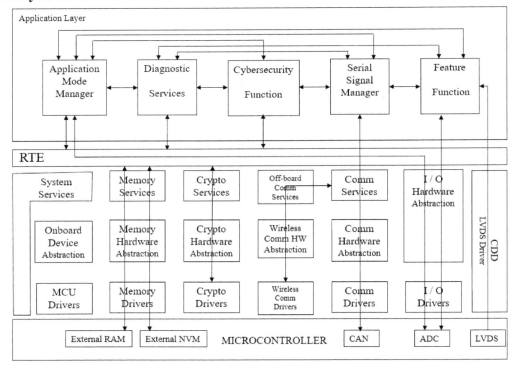

Figure 3.4-30 Data Flow in Structure

To derive the safety requirement for the infrastructure components including the autosar and the bridge software components introduced in the section of 3.4.2 System Structure Design, the data flow analysis in the entire ECU should be done, which is described in Figure 3.4-30 Data Flow in Structure including:

- The CDD is ASIL B

- The I/O HW abstraction and driver are QM
- The comm components are ASIL B
- The memory components are ASIL B
- The system service, onboard device abstraction and MCU drivers are ASIL B
- The crypto components are ASIL B because they provide the authenticated communication

In the application layer:

- The feature function component is ASIL B because it provides the object detection.
- The serial signal manager component is ASIL B because it provides the enabling signals.
- The Cybersecurity function is ASIL B because it provides the authenticated communication including the enabling signals.
- The diagnostic services component is QM because it does not impact any safety goals.

The application mode manager is ASIL B because it interacts with all the application layer components above, among them there are the ASIL B components.

Reliability Measurement:

Based on the analysis above and the approach described in the section of 3.4.5 FMEA, the reliability measurements will be:

- LVDS data flow specific safety requirements:

LVDS Safety Requirement 1: The LVDS driver shall detect the transmission faults using the Line-Fault indication.

LVDS Safety Requirement 2: The LVDS driver shall detect the transmission clock faults.

- CAN data flow specific safety requirements:

CAN Safety Requirement 1: The CAN driver shall detect the message value errors by checking the messages' CRC.

CAN Safety Requirement 2: The Serial Signal Manager shall detect the signal value errors by checking the checksum value.

CAN Safety Requirement 3: The Serial Signal Manager shall protect the message repetition errors by checking the message counter value.

CAN Safety Requirement 4: The Serial Signal Manager shall protect the message from the transmission timing errors.

- ADC data flow specific safety requirements:

ADC Safety Requirement 1: The I/O driver shall ensure that the ADC is configured correctly.

Safety Requirement 2: The Application Mode Manager shall prevent the ADC configuration from unintended changing during the runtime.

- External memory data flow specific safety requirements:

EM Safety Requirement 1: Both external RAM and NVM shall prevent them from the permanent damage by marking and avoiding the accessing to the damaged aera.

- Power Supply Unit specific safety requirements:

PW Safety Requirement 1: The power supply unit shall be configured to use the self-detect the voltage out of range.

- RPU Cores specific safety requirements:

RPU Safety Requirement 1: The RPU0 and RPU1 shall be configured to use the "Lock-Step" mode.

- For RPU, GPU and APU Cores specific safety requirements:

Core Safety Requirement 1: The Platform Management Unit (PMU) including the memory management unit (MMU), peripheral management unit (PMU) shall be used to monitor the microcontroller.

Core Safety Requirement 2: The Watchdog shall be used to monitor the PMU, AUTOSAR and application components to prevent them from execution errors.

Core Safety Requirement 3: All clocks shall be monitored to prevent the processors from the time errors.

Core Safety Requirement 4: The Built in Self-Test (BIST) and Error Correction Code (ECC) measures shall be used for the RAM, ROM, communication bus control register, logic control register, interrupt control register, etc.

Availability Measurement:

To increase the availability, the following measurements will be implemented

- Output warning signals

Availability Safety Requirement 1: The FPGA shall implement the backup output warning signal mechanism.

- External NVM data

Availability Safety Requirement 2: The AUTOSAR NVM data block backup functions should be configured.

- Recovery mechanisms
  - Communication devices

Availability Safety Requirement 3: The CAN transceiver shall be reset if the CAN bus off error is detected.

Availability Safety Requirement 4: The CAN transceiver shall try to reset again until the reset threshold value (3) is reached if the reset is failed.

Availability Safety Requirement 5: The Deserializer shall reset if the errors are detected.

Availability Safety Requirement 6: The Deserializer shall try to reset again until the reset threshold value (3) is reached if the reset is failed.

  - External NVM location change

Availability Safety Requirement 7: The NVM manager shall change the often-used parameters' location in the NVM if the location access counter number reaches 95% of the NVM lift time value.

  - ADC

Availability Safety Requirement 8: The application mode manager shall reset the ADC if the ADC reports any device error.

  - Partition

Availability Safety Requirement 9: The partition shall be reset if the following errors are detected in the partition:

- Memory access violation
- Watchdog Timeout
  - Core

Availability Safety Requirement 10: The core shall be reset if the following errors are detected in the core:

- Memory access violation in the master partition
- Memory and peripheral errors reported from the PMU in the master partition
- Watchdog Timeout in the master partition
  - ECU

Availability Safety Requirement 11: The ECU shall be reset if the Application Mode Manager requests the reset.

Quality Control Measurement:

The quality control is based on the experience to sum up the best practice to do certain activities, and the best and the most efficient way is to establish the suitable processes at the organization level to be compliant with the industrial standards: ASPICE, ISO 26262, which is the project management and out of the book scope.

In addition to the development processes, there are following common software execution control activities that can be done to increase the software execution reliability from the technical point of view, and the contents of which are not to detect errors, so they are kind of "dynamic quality control" in the development though the execution monitor using the watchdogs is overlapped with the reliability development.

State Control:

In the control running life cycle, there are different states, for example, in Figure 3.4-12 State Flow in Mode Manager, there are 6 states in an automotive ECU running life cycle, in each of them the ECU functionality which is represented by the data has different "heath status":

- RUN State: which is the normal state, in which all functions are running as implemented.
- Power Off State: All functions are stopped except the wake-up function.
- Sleep State: Only the designed functions are running in the low power mode.
- Start Up States: In which, all devices are powered up, and all data are initialized, so all data in the state are in the "un-trusted" state before the initialization is done, and usage of data may cause some issues.
- Wake Up State: In which, all slept devices are powered up, and all data in the slept functions are initialized, so all data in those slept devices and functions are in the "un-trusted" state before the initialization is done, and usage of any those data may cause some issues.
- Shut Down State: Since the shutdown may be caused by fault detection, so some functionalities may not be stable, so the usage of data needs to be carefully designed.

In system development, usually the "critical" points must be paid attention, which are the points or states, in which the data are not certain, and most of those critical points are allocated on the "boundary" states that are Start Up, Wake Up and Shut Down in the ECU running life cycle above.

The data, functions and devices have the following dependency relationship and characteristics, which have different value or state in different execution cycle phase:

- the data are from the functions, such as:
    - locate defined data
    - static defined data
    - global defined data
    - data initial value is defined by the compiler
    - data initial value is defined by the application

- the functions are existed in the partitions or provided by the devices, which depend on the state where they are allocated.
- the partitions are existed in the processors, which need the following activities:
  o Configure the partitions
  o Load the needed operating system and other services components into the partitions
  o Initialize all component in the partitions to their normal states
- the devices are controlled by the processors, the activities of which consist:
  o Load the pre-defined configurations into the devices' control registers.
  o Execute the built-in self-test (BIST)
  o Execute the application specific test if needed

So, when the ECU is in the critical states, such as:

- In Power Up or Wake Up state, the devices and functions need to be initialized, configured, tested.
- In Shutdown state, some functions need to be disabled.
- In the Fault state, some data values are not trustable

All those situations will impact the data status and availability, which should be designed carefully.

The Application Mode Manager in the section of 3.4.2.2 Application Mode Manager is the component to synchronize all the application layer components including itself to initialize and shut down them by collaborating with the EcuM in the AUTOSAR.

Once the ECU is in the RUN State, the following measures can be done to reduce the errors that includes both the mistakes and the systematic errors.

- Data range check
- Execution monitor

The practical approach to do the data range check is based on the FMEA result, which is that the data in the high-risk data flow will be check against the design data range at time:

- Before the data are taken into the calculations.
- Before the data are output to the external functions.

Execution Monitor:

The watchdog can be used to monitor if the execution follows the design logic, which consists:

- Internal Watchdog:
  o Sigle Point Watchdog: which is to reset the watchdog timer at certain points in the execution flow by design, if the execution does not follow design logic to reset the watchdog timer, then the watchdog will timeout to generate the interrupt to inform the controller about the error.
  o Window Watchdog: which is to set the time window using two-time values: top value of the window and bottom value of the window to monitor certain piece of code execution, if the execution exceeds the time window, then the watchdog will generate the interrupt about the error.
- External Watchdog: which is a redundant safety measure to monitor the execution to detect if the entire microcontroller is out of control, in which case the microcontroller internal safety mechanisms may not be trustable, so the external

watchdog will reset the microcontroller to recover.

Verification:

Verification is a common sense and is the engineering principle: the work product is not trustable if it cannot be verified.

So, verification should be done as the quality control measure in every development step, especially at the milestones and to every work product.

The detailed verification will be introduced in the section of 3.5 System Verification.

3.4.6.4   ISO 26262 Compliance

In the automotive industry, the development and lifecycle of electrical and/or electronic safety relevant systems are required to comply with ISO 26262 standard: Road vehicles – Functional safety, which is a very detailed and complete standard about the automotive electrical and/or electronic systems development, although its name is "Functional Safety", it covers all the necessary steps in the whole development lifecycle. To be compliant with ISO 26262, the ECU developers need to demonstrate two aspects:

- Good technology: the developers need to demonstrate that the product design and implementation good enough to meet the required safety requirements.
- Good faith: the developers need to demonstrate that the development is managed by the good processes that are required by ISO 26262.

Terminology Clarification:

The advantages of ISO 26262 are that it is a very detailed and complete standard about the automotive electrical and/or electronic system developments by covering all the necessary steps in the whole development lifecycle, which are applicable to not only safety-related systems but also the non-safety systems.

However, one of the disadvantages of the standard is: there are too many terminologies that are confusing and non-value added.

One of most confusing aspect is in the development steps, especially in the "Concept Phase" in the Part 3 of the standard, in which, there are the terminologies of technical safety concept, technical safety requirement, functional safety concept, functional safety requirement, all of which are not necessary because the item attributes or the meanings that the item represents can be easily distinguished by the item location and from the item environment.

In every product development, such as a vehicle, an assembly, an ECU, a component or a function, the development procedure is always the three steps:

- First step: do the requirement analysis to derive the requirement specifications.
- Second step: design the product according to the requirements, which either derives the requirement specifications for the next level development or achieves the final required product.
- Third step: verify the results in the second step against the specification from the first step to check if the designs comply with the requirements, which is either the specification verification or product verification.

The same development procedures are specified in the standard including the "Concept Phase". but it may be confusing due to the terminology. For example, the following steps in the standard follow development procedures above, but the special terminologies are used:

In the Part3 "Concept Phase":

- Requirement Analysis: which is called as the Item definition
- Design: This is the hazard analysis and risk assessment, which derives the Functional Safety Concept for the next level development.

In the Part 4 "Product development at the system level"

- Requirement Analysis: This is the functional safety concept from the Part 3 above.
- Design: This is the technical safety concept including the system elements and interfaces, which derives the system architectural design for the next level development.

In the Part 5 "Product development at the hardware level"

- Requirement Analysis: This is the system architectural design from the Part 4 above.
- Design: This is the hardware design including the hardware-software interface (HSI) specifications, which derives the hardware implementation specifications.

In the Part 6 "Product development at the software level"

- Requirement Analysis: This is the system architectural design from the Part 4 above.
- Design: This is the software architectural design, which derives the software unit design for the software unit development

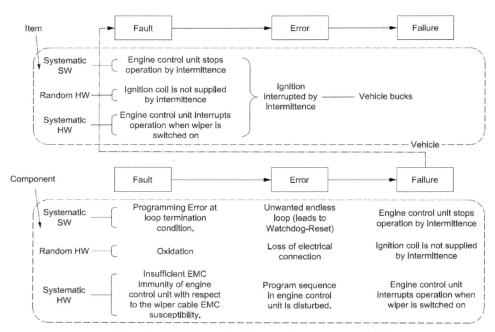

Figure 3.4-31 Relationship between Fault, Error and Failure

So, it will be clearer and more direct to indicate an item development by using the requirement, design and verification, in which, if the item requirements are safety relevant, then the design result including decompositions and other derivations will be safety relevant, as well, based on the derivations which can follow the roles specified in the Part

9 of the standard. If the development needs to distinguish the development levels, then it can indicate that the development including the requirement engineering. design and verification is at the concept level, system level or software level. One example of such distinguishes is the network model defined by the levels of physical level, electronic level, data link level, protocol level, application level, etc.

Another confusing terminology aspect in ISO 26262 is that the terms of Fault, Error, Failure and Failure Mode are distinguished, for example, the relationship between the Fault, Error and Failure is illustrated in Figure 3.4-31 Relationship between Fault, Error and Failure below, which is from the Part 9 of ISO 26262.

However, in this book, all those terms have the exact same meaning: something is wrong, which means: "The result does not follow the requirement(s)". The rationales are:

- The distinguishing them does not add any value to the development, rather, it leads easily to confusing.
- The errors at the different level can be easily identified by the context, such as the systematic SW Fault in Figure 3.4-31 Relationship between Fault, Error and Failure below:
  - SW programming error at the loop termination condition in the Engine Control ECU (which is clearly described about what is wrong and where).
  - "Engine control unit stops operation by intermittence" can be described equally as:
    - The engine control unit has a failure of stopping operation by intermittence
    - The engine control unit has an error of stopping operation by intermittence
    - The engine control unit has a fault of stopping operation by intermittence
    - The engine control unit has a failure mode of stopping operation by intermittence

The relationship in Figure 3.4-31 Relationship between Fault, Error and Failure above can be clearly described as:

The Engine Control ECU has a SW programming error at the loop termination condition, which leads to the error that the Engine control unit stops operation by intermittence, which leads to the error that the ignition is interrupted by intermittence, which leads to the error that the vehicle bucks.

Development technology Compliancy

The ISO 26262 provides the quantitative measures about what criteria should be met in the development. The first measurement is the product ASIL rating consisting of QM, ASIL A, ASIL B, ASIL C and ASIL D listed in Table 3.4-14 ASIL Ratings below:

Based on the combinations above, the standard defines following Automotive Safety Integrity Level (ASIL) to classify the safety goals and safety requirements:

- ASIL D is the highest level,
- ASIL C is the second highest level,
- ASIL B is the third highest level,
- ASIL A is the fourth highest level,
- QM is the lowest level that is for safety irrelative products.

Severity Class	Probability Class	Controllability Class		
		C1	C2	C3
S1	E1	QM	QM	QM
	E2	QM	QM	QM
	E3	QM	QM	A
	E4	QM	A	B
S2	E1	QM	QM	QM
	E2	QM	QM	A
	E3	QM	A	B
	E4	A	B	C
S3	E1	QM	QM	A
	E2	QM	A	B
	E3	A	B	C
	E4	B	C	D

Table 3.4-14 ASIL Ratings

To achieve certain ASIL or to development required ASIL product, the standard defines certain activities and processes must be done in the development.

Target Metric	ASIL A	ASIL B	ASIL C	ASIL D
SPFM (Single-Point Fault Metric) (Table 4 in ISO 26262 Part5)	Not Required	>=90%	>97%	>=99%
LFM (Latent Fault Metric) (Table 5 in ISO 26262 Part5)	Not Required	>=60%	>=80%	>=90%
PMHF (Probabilistic Metrics for Hardware Failures) (Table 6 in ISO 26262 Part5)	Not Required	<=100FIT	<=100FIT	<=10FIT

Table 3.4-15 Hardware Fault Metrics

For the hardware development, the quantitative criteria are in Table 3.4-15 Hardware Fault Metrics. Among the table:

LFM: Latent faults are multiple-point faults whose presence are not detected by a safety mechanism nor perceived by the driver within the multiple-point fault detection interval. The latent fault metric (LFM) is a hardware architectural metric that reveals if the coverage is sufficient by the safety mechanisms to prevent risk from latent faults in the hardware architecture.

SPFM: Single point faults are faults in an element that are not covered by a safety mechanism and that lead directly to the violation of a safety goal. The single point fault metric (SPFM) is a hardware architectural metric that reveals if the coverage is sufficient by the safety mechanisms to prevent risk from single point faults in the hardware architecture.

PMHF: Probabilistic metric for random hardware failures (PMHF) is expressed in FITs.

FIT: The number of failures that can be expected in one billion ($1x10^9$) device-hours of operation, that is: one fault in 1,000,000,000 hours.

For the software development, there are not the quantitative criteria for software execution because the software in a computer system must always execute in the same way as implemented except the error that are cause by platform hardware.

To regulate the software development, there are the required development aspects for the different ASIL rating products, although there is not any clearly defined quantitative measurements.

Safety issue in autonomous driving based on ISO 26262

Based on the safety criteria set in Table 3.4-15 Hardware Fault Metrics. the safety of autonomous driving systems is quite questionable.

SAE Level	Automation Level	Definition
Human driver monitors the driving environment		
0	No Automation	The human drivers do the driving all the time.
1	Driver Assistance	The human drivers do the driving most of time except some cases where the Driver Assistance systems provide some actions, such as warnings, emergence brake or steering.
2	Partial Automation	The human drivers do the driving mainly except some cases where the Automation systems provide some actions, such as automatic parking.
Automated driving system monitors the driving environment		
3	Conditional Automation	The Automation systems do some daily driving in some cases, such as on the highway, and the human drivers will respond appropriately to a request to intervene.
4	High Automation	The Automation systems do most of driving, and the human drivers will only need to respond in some cases where the driving is complicate.
5	Full Automation	The Automation systems do all driving where as long as the human drivers are capable of.

Table 3.4-16 SAE Autonomous Level

There are not the dedicated criteria for the autonomous driving system, there only the definitions listed in the Table 3.4-16 SAE Autonomous Level. However, the hardware safety criteria above can be the reference, because from a black box point of view or from the system external functionality point of view, there is not any difference between a hardware component and a driving subsystem.

Every automation system used in any level above must be the ASIL D except the level 0. And currently the object detection systems used in the autonomous driving system either level 1 or level 2 have the confidence rate: 95% ~ 97%, which is: the best case that the system can detection the required objects is 97 correct cases out of 100 detections. Even if the object detection system could reach the 99% confidence rate, then that is 1 mistake out of 100 detections, and it is reasonable to assume that a vehicle will meet 100 objects in an hour, then comparing which with the PMHF required by the ASIL B in hardware criteria, that will be 10 million times difference, i.e., the current object detection system must improve 10 million times to meet the ASIL B criteria.

Development Process Compliance

The ISO 26262 defines the specific processes in the whole product development life cycle including the system operation concept design, system architecture design, hardware design, software design, verification and test, the activities of those development are allocated the following levels:

- Concept Level
- System Level
- Hardware Level
- Software Level
- Verification

Concept Level

The Concept Level development is described in the Part3 "Concept Phase" including:

- Item definition
- Hazard analysis
- Functional safety concept design

In which, the development needs to do the item definition including the item safety requirements and item non-safety requirements, then based on which, the development needs to do the functional safety concept design that is based on the hazard analysis that is based on the item definition.

If the developers try to do this "Concept Phase" as the normal Top-Down system development, then the concept phase is the most confusing part in the standard, the issue is the step of hazard analysis, which should be either part of item definition that is defined by the higher-level development or the phase after the item functional structure design, because:

- if the hazards are caused by the item internal malfunctions, i.e., the hazard are the effects of the item failures, then the hazard analysis should be done the higher-level product design;
- if the hazards cause the item malfunctions, i.e., the hazards are the failure causes to the item effects, then the hazards are the item internal malfunctions which must be done based on the item structure. In fact, the standard indeed requires the "System architecture design (from external source)" as the

prerequisite, but the prerequisite is for the Functional Safety Concept, not for the hazard analysis.

It would be much easy to understand the "Concept Phase" in this way:
- The first step of "Concept Phase" is to do both of following:
  - the system requirement elicitation and engineering which are required in the ASPICE.
  - the system operation concept design which expresses the required system using the input, output and middle data and the relationships between the data which is described in 3.4.1 System Operation Concept Design
- The second step of "Concept Phase" is to design the system reliability or "error tolerate" or "fail-safe" mechanisms based on the failure and effect analysis.

The reason why the confusions exist is because that the standard focuses on only the "fail-safe" mechanism development, so it separates the safety development from the normal system development, which makes the development confusing and somehow duplicate. So, the suggestions from the book are to do all the developments required by the standards together with the normal system developments because the "fail-safe" is part of the system under development, and there is no added value to separate them.

In the Data Driven system engineering of this book, the concept phase relevant activities are:
- Item Definition: the item definition is covered by the sections of the requirement elicitation, the requirement engineering and the system operation concept design. In the sections of 3.2 Requirement Elicitation and 3.3 Requirement Engineering, the full BSD system definition are given by the requirements, in which, the left camera BSD system is an ASIL B (D) product, and the output signal: CAN_BSD_Left_Alert (TTC) is an ASIL B signal, the hardware that supports the CAN_BSD_Left_Alert (TTC) implementation must be the ASIL D. Those safety requirements are inherited from the higher-level components, which is described in "Safety at the assembly level" of 3.4.6.3 BSD Safety.
- Hazard Analysis: the hazard analysis is covered by the section of 3.4.5.2 Data Driven FMEA, in which, the full potential failure modes are described using the data driven approach, in which, according to the characteristics in a computer system that the system functionality can be fully represented by the two attributes from each data in the system: Data Value and Data Timing, and the data in the system are interacted each other according the data derivation formulas, based on which, the full system failure mode can be figured out.
- Functional Safety Concept: the functional safety concept is done by the combinations of system operation concept, system structure design, FMEA and verification. First, based on the failure mode analysis from the Data Driven FMEA that is based on the system operation concept, the safety mechanisms against the failure modes are design and allocated to the suitable components based on the system structure, then those mechanisms will be verified in the verification phase, which mainly focuses on the system integration test described in 3.5.2 System Integration and Verification. In the table of Table

3.4-12 BSD DD FMEA Analysis, the safety mechanisms are designed as preventive actions and the detection actions at the system architectural level.

The key points to meet the development requirement in the concept phase:
- What are the requirements, especially the safety requirements?
- What faults may impact the safety requirements above?
- What protections should be developed to prevent the product from the faults above?

System Level
The System Level development is described in the Part 4 "Product development at the system level", in which, the development needs to do the system architectural design including the system elements and their interfaces, especially the development needs to implement the safety mechanism to prevent the system from the potential errors. From which the system element definitions and the interface definitions that are for both system functionality and the safety are available for the next level development that are the hardware and software development.

The safety development at the system level above can be described in the easy way using the three-step procedure that is introduced at the beginning of this section:
- Requirement: Functional safety concept and the system architectural design

Which are the input information from the concept phase to provide the safety requirements from the concept point of view.
- Design: the technical safety concept and the system architecture design.

In the Data Driven system engineering of this book, the system architecture design is covered by the section of 3.4.2 System Structure Design, in which, the application layer components: Feature Function, Application Mode Manager, Serial Signal Manger, Cybersecurity Function are defined together with their interfaces, and the integration of AUTOSAR.

Meanwhile, the contents of "Reliability Measurement" and "Availability Measurement" in 3.4.6.3 BSD Safety need to be considered in the system architecture, and the measures in the two sections above need to be allocated to the system structural components, for example, the two analog signal handlings should be allocated to the ADC device and the I / O driver or other software component depending on the system structure:

Environment date flow:
Veh_Power = f_power (
ADC_Veh_Power,
Veh_Power_Filter_Constant,
ADC_Veh_Power_Validity);

Env_Temperature = f_temp (
ADC_ Env_Temperature,
Env_Temperature_Filter_Constant,
ADC_Env_Temperature_Validity);

And to implement the safety mechanisms for the middle data flow below:
Object_Detected   (Object_Relative_Position,   Object_Relative_Velocity)   =
f_detection();

The relevant safety mechanisms related to Object_Detected, Object_Relative_Position and Object_Relative_Velocity need to be designed, according to the Table 3.4-12 BSD DD FMEA Analysis, the detection measures of Value Range Check, Reasonable Value Check need to be considered and to be allocated to the responsible components, and the allocated measures will be implemented by the hardware and software development.

One of key work products in this part is the Hardware Software Interface (HSI) design, the reason for which is that the most of automotive ECUs, especially the conventional ECUs, are for motion control, for example, the braking control, the steering control, the driver and passenger doors control, the power liftgate control, which need to output the control signals to the actuators, so the interfaces to those actuators should be carefully design and configured, and calibrated during the operation, so that the motion control can be accurate and in time. Meanwhile, to perform those motion control actions, the vehicle needs to measure the vehicle dynamic status using either analog or digital signals, and to read those signals, the interfaces for the measurement devices should be defined.

However, for some ECUs, such as the left camera BSD ECU, it does not have any special device, all the I/O devices, such as CAN, LVDS, are industrial standard devices, the HSI for this kind of ECUs should be design as the normal functional components.
- Verification: the tests and verifications against each work products above.
Verifications will be described in the section of 3.5 System Verification.

The key points to meet the safety requirement in the system level development:
- To meet the requirements, especially the safety requirements from the concept phase, what system structure including the system elements should be designed, and how do the system elements interact each other to fulfill the requirements?
- What are the requirements from the system level to the next level development?
- How to verify the development at the system level?

Hardware Level
The Hardware Level development is described in the Part 5 "Product development at the hardware level", in which, the development needs to do the hardware architectural design including the hardware elements and the Hardware - Software interfaces (HSI).

The hardware development is not covered in detail by this book, only the key hardware design points are covered in the section of 3.4.3 Electronic Architect Design and some very high-level examples are given in the BSD example, in which, the "Safety at the processor level", the "Reliability Measurement" and "Availability Measurement" sections in 3.4.6.3 BSD Safety need to be considered in the hardware architecture, and the measures in those sections above need to be allocated to the hardware structural components, such as RPU, APU, etc., and the communications between those hardware components need to be design.

However, based on ISO 26262, there are following critical points about hardware development:
- All the safety relevant hardware components must meet the required safety ratings (LFM, SPFM, PMHF) in the Table 3.4-15 Hardware Fault Metrics.
- For each safety relevant hardware device that does not meet the required safety rating mentioned above, then the corresponding safety enhancements to reach the

required safety rating must be designed, such as invoke the built-in self-test routines cyclically.

- If a hardware safety requirement is derived from the safety requirement decomposition, then the hardware safety requirement must inherit the highest safety rating from the parent requirement, for example, in BSD system that is a ASIL B(D) system, so the hardware must be ASIL D though the system is ASIL B.
- If the safety relevant hardware devices are not the industrial standard devices, then those device interfaces must be clearly defined, and which will be done in the CDD if the AUTOSAR is used.

The key points to meet the safety requirement in the hardware level development:
- To meet the requirements, especially the safety requirements from the system level, what hardware structure including the hardware elements should be designed, and how do the hardware elements interact each other to fulfill the requirements?
- Do the hardware components meet the requirements listed in Table 3.4-15 Hardware Fault Metrics?
- How to verify the development at the hardware level?

Software Level

The safety mechanism development at the software level is described in the Part 6 "Product development at the software level", in which, the software safety mechanisms need to be derived from the software safety requirements that are from the system level development and allocated to the software by the software architectural design including the software elements and the software unit detail design, to do which, the three development steps will be used:

- Requirement: which are the safety requirements allocated to the software components from the system safety development.
- Design: During the software safety mechanism design, the individual and specific safety measure will be derived to and allocated to the responsible software component, and then the execution and interaction mechanism between those components, and with other non-safety software components including the interactions with the hardware devices should be designed. The design procedure is same as the one in the system development level, the only difference is that the software safety development is only allocated at the software functions, which is same for the software detailed development including the software detailed design and implementation, i.e., the only difference between the software safety development at the software level and the software safety development at the software unit level is that at the software unit level development needs to derived and allocated the safety mechanism to the more detailed software components, such as the detailed software functions that can be implemented using the programming languages.

Based on the AUTOSAR structure, if the safety relevant hardware devices are not the industrial standard devices, then those device interfaces must be carefully defined using the CDD component?

If the software components with different ASIL rating exist in a processor, then the Freedom From Interference (FFI) mechanism must be implemented, for example, the different ASIL rating software components should be separated using the partitions.

If the communications exist between the components with different ASIL rating in the ECU, then the Freedom From Interference (FFI) protection mechanism should be implemented by the software, i.e., the software component with the higher ASIL rating must check the information received from the component with lower ASIL rating.

- Verification: The systematic verification including the software verification will be described in the section of 3.5 System Verification, and in principle, there is not significant difference between the system verification and the software verification. However, since the software implementation and integration are hardware platform specific, i.e., the APU software components are all in the same environment, so the software verification can be done more easily than the system verification by using the Integrated Development Environment (IDE) or other simulated environment.

The software development is not covered in detail by this book, as mentioned at the beginning of this book, there is not any significant difference between the system and the software development, so all the development activities described at the system level in the book are applicable for both hardware and software levels development.

In the section of 3.4.6.3 BSD Safety, some example of software safety mechanism are designed to ensure the BSD system safety, in which, the "Safety at the partition level" and the "Safety at the function level" together with the "Reliability Measurement" and the "Availability Measurement" sections in 3.4.6.3 BSD Safety need to be considered in the software architecture design, and the safety measures in those sections above need to be allocated to the software structural components, for example, the detection measures of Value Range Check, Reasonable Value Check in Table 3.4-12 BSD DD FMEA Analysis need to be considered and to be allocated to the responsible software components, and the measures of dynamic software status control, such as State Control, Execution Monitor in 3.4.6.3 BSD Safety components need to be designed.

The key points to meet the safety requirement in the software level development:
- To meet the requirements, especially the safety requirements from the system level, what software structure including the software functions should be designed, and how do the software functions interact each other and interact with the hardware elements to fulfill the requirements?
- Does the software development comply with the safety requirements: ASIL ratings, FFI?
- How to verify the development at the software level?

Verification

In ISO 26262, the verification should be done at every step of every phase: the functional safety concept, the technical safety concept, the system architectural design, the hardware architectural design, the software architectural design, the software unit design and implementation.

3.4.6 Safety

The verification in this book is described in the section of 3.5 System Verification, which is applicable also to both software and hardware detailed design and implementation verifications.

The verification described in the Part 8 of the standard focuses on the development process verification and development environment verification, which is out of this book's scope.

The key points to meet the safety requirement in the verification:

- Does the verification cover all functionalities?
- Does the verification have enough granularity?
- Does the verification cover all aspects in the development?

Check List 8 – Safety

- Generic (applicable for each phase)
    - o Do the development quality control processes including the layered development and verification follow ISO 26262?
    - o Do the safety requirement decompositions including the safety rating inheritance and allocation follow ISO 26262?
- Safety Concept
    - o Are the system safety requirements clearly derived from the vehicle safety requirements?
    - o Are the system performance limitations classified regarding with vehicle sensing and acting abilities to both road and road users?
    - o Are the SOTIF misuse preventions strategies defined?
    - o Is the system operation concept or system architectural design available that should have all the derivations of system output signals?
    - o Are the system safety requirements derived to the responsible system elements in the system operation concept or system architectural design?
    - o Are the system safety mechanisms designed based on the safety requirements by covering reliability, functional availability and development quality?
    - o Do the reliability mechanisms cover all potential error detections including the data value error and data timing error?
    - o Do the error detections cover the input data, middle data and output data about value error and timing error?
    - o How are the middle data value errors detected (range check, reasonable value comparison, comparison between duplication)?
    - o How are the middle data timing error protected (watchdog, window watchdog, external watchdog, scheduler design, execution path check)?
    - o What are the output signal safety protections implemented (integrity, time)?
    - o Are the safe states designed for each fault that impacts the system safety?
    - o Do the FTTIs of safe states meet the system safety requirements?
    - o Do the safe state actions (such as communication silence, warning, device reset) protect the system safety?
    - o Are the recovery mechanisms from each fault state designed?
    - o Are the functional availability mechanisms designed according to the ISO26262, so that the duplicated system elements are independent each other, and each of them can fulfill the required system safety independently?
- Safety development at system level
    - o Are the system safety requirements clearly derived to the responsible system elements (Feature Function, Application Mode Manager, Serial Signal Manger, Cybersecurity Function, AUTOSAR)?
    - o Are the safety requirements that are allocated to the responsible system elements clearly derived to the responsible software and hardware components?
    - o If there are the safety requirements allocated to the devices rather than the industrial standard devices, are the interfaces between those devices and

their software clearly defined?

- o For each safety relevant system element, are the corresponding safety mechanisms designed?
- o For each safety relevant system element, are the corresponding functional availability mechanisms designed if needed?
- o Is every safety measure verified with the corresponding testing granularity?
- Safety development at hardware level
  - o Do all the safety relevant hardware components meet the required safety ratings (LFM, SPFM, PMHF)?
  - o If a hardware safety requirement is derived from the safety requirement decomposition, then does the safety requirement inherit the highest safety rating from the parent requirement?
  - o Are the safety requirements that are allocated to the responsible hardware elements clearly derived to the responsible sub-level hardware components if needed?
  - o If the safety relevant hardware devices are not the industrial standard devices, are those device interfaces clearly defined (by the CDD if the AUTOSAR is used)?
  - o For each safety relevant hardware device that does not meet the required safety rating, are the corresponding safety enhancements to reach the required safety rating designed?
  - o For each safety relevant hardware element, are the corresponding functional availability mechanisms designed if needed?
  - o Is every safety hardware element verified with the corresponding testing granularity?
- Safety development at software level
  - o Are the safety mechanisms that are allocated to the responsible software elements designed at the software level?
  - o Are the safety requirements that are allocated to the responsible software elements clearly derived to the responsible sub-level software components if needed?
  - o If the safety relevant hardware devices are not the industrial standard devices, are those device interfaces clearly designed (by using the CDD if the AUTOSAR is used)?
  - o If the software components with different ASIL rating exist in a processor, is the Freedom From Interference (FFI) mechanism implemented?
  - o If the communications exist between the software components with different ASIL rating in the ECU, is the Freedom From Interference (FFI) protection mechanism implemented in the higher ASIL rating software component that receives signals from the lower ASIL rating software component?
  - o For each safety relevant software element, are the corresponding functional availability mechanisms designed if needed?
  - o Is every safety software element verified with the corresponding testing granularity?

3.4.7 Cybersecurity

In this section, this book provides the comprehensive analysis and solutions for the cybersecurity implementations in the ECUs that are behind the Gateway ECU and allocated in the strictly managed local vehicle network illustrated in Figure 3.4-32 Vehicle Network Structure, which are the ones indicated by ECU -A1~An, ECU-B1~Bm, ECU-G1~Gk, for the other ECUs, such as the Gateway ECU, GPS ECU, CarPlay ECU, wireless remote control receiver ECU like Key Forb ECU, are not the focus of this book.

Due to the impact to the aspects of safety, finance and privacy, the vehicle cybersecurity requirements become critical, in some countries, such as Europe, North America, South Kora and Japan, the cybersecurity requirements specified in the UN ECE 155 / 156 are mandatory.

The network characteristics in a vehicle illustrated in Figure 3.4-32 Vehicle Network Structure below leads to the significant different security & privacy requirements to the automotive ECUs comparing with the daily used computers that are connect to the internet, among which, the main differences are:

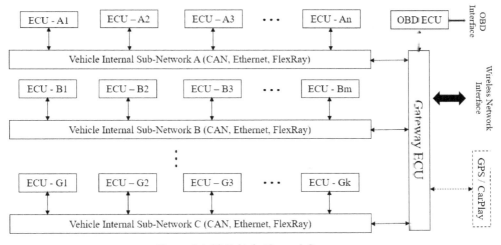

Figure 3.4-32 Vehicle Network Structure

- Execution Contents: all contents in the automotive ECUs are extremely controlled including the updating and modifying, however, for the daily used computers, the updates or download contents from various websites happen time to time.
- Communication: The communications between the automotive ECUs inside of the vehicle network behand the Gateway ECU illustrated in Figure 3.4-32 Vehicle Network Structure above are pre-defined, such as which ECUs should send what messages to which ECUs. However, for the daily used computers, they may go to any websites on the internet to interact with any software.

So, in a vehicle network, only the Gateway ECU and On-Board Diagnostic (OBD) Interface ECU are similar as the daily used computers from outside of vehicle point of view, those ECUS need to implement the protections about the threats, vulnerabilities and attacks as the daily used computers do, which are out of this books' scope. All other ECUs that are behind the Gateway ECU and allocated in the strictly managed local vehicle network, which are the focus in this book.

For the ECUs, such as GPS ECU, CarPlay ECU, wireless remote control receiver ECU like Key Forb ECU, they need to connect to outsides, there are two options:
- Those ECUs can be allocated inside of the vehicle network behand the Gateway ECU, which can make use the security functions that are developed in the Gateway ECU, however, that will increase the communication load on the Gateway which will impact the performance for both the vehicle network and those ECUs.
- Those ECUs can be allocated outside of the vehicle network before the Gateway ECU as illustrated in Figure 3.4-32 Vehicle Network Structure, so that those ECU will not interfere the ECUs inside of the vehicle network, then increase the security for them, however, those ECUs must implement the full security functions to protect themselves like the daily used computers, which are out of this book's scope.

3.4.7.1 Cybersecurity Threat Impact

The cybersecurity threat impact from the automotive ECUs is based on the characteristics described in the section of 2.2.1 Basic Structure, especially based on the vehicle network structure illustrated in Figure 3.4-32 Vehicle Network Structure, which consists of Safety Impact, Financial Impact and Privacy Impact.

In the ISO 21434, there is the operational impact from the cybersecurity threat, which, from the author point of view, can be classified into either safety or financial impact, so the operational impact is a duplicated point and it will not be discussed in this book.

- Safety Impact: A vehicle average weight is about a couple of tons, if it is out of the driver control, then it would become very dangerous object, especially when it is running at high speed or running in the aera where there are people around. The safety impact ratings and the criteria are listed below from the ISO 21434:

Impact rating	Criteria for safety impact rating
Severe	S3: Life-threatening injuries (survival uncertain), fatal injuries
Major	S2: Severe and life-threatening injuries (survival probable)
Moderate	S1: Light and moderate injuries
Negligible	S0: No injuries

Table 3.4-17 Safety Impact

- Financial Impact: A vehicle is the second most expensive asset in people's life after the house, anything that is out operation or damaged will be financial impact. And if someone can use some "pay to use" functions without the payment, then it would be the dealers or OEMs financial lost. The financial impact ratings and the criteria are listed below from the ISO 21434:

Impact rating	Criteria for financial impact rating
Severe	The financial damage leads to catastrophic consequences which the affected road user might not overcome.
Major	The financial damage leads to substantial consequences which the affected road user will be able to overcome.
Moderate	The financial damage leads to inconvenient consequences which the affected road user will be able to overcome with limited resources.
Negligible	The financial damage leads to no effect, negligible consequences or is irrelevant to the road user.

Table 3.4-18 Financial Impact

- Privacy Impact: Nowadays, there are more and more functions in vehicles to connect the cell phones to provide certain convenient functions, such as radio control, play music, make sure the cell phone applications, in which, some persons' private information will be accessed, such as person ID, location, credit card information. If that information is accessed by un-authorized persons, then it would impact people's life, which may result in some legal actions. The privacy impact ratings and the criteria are listed in Table 3.4-19 Privacy Impact below from the ISO 21434:

Impact rating	Criteria for privacy impact rating
Severe	The privacy damage leads to significant or even irreversible impact to the road user. The information regarding the road user is highly sensitive and easy to link to a PII principal.
Major	The privacy damage leads to serious impact to the road user. The information regarding the road user is: a) highly sensitive and difficult to link to a PII principal; or b) sensitive and easy to link to a PII principal.
Moderate	The privacy damage leads to inconvenient consequences to the road user. The information regarding the road user is: a) sensitive but difficult to link to a PII principal; or b) not sensitive but easy to link to a PII principal.
Negligible	The privacy damage leads to no effect or, negligible consequences or is irrelevant to the road user. The information regarding the road user is not sensitive and difficult to link to a PII principal.

Table 3.4-19 Privacy Impact

In the Annex 5 of UN ECE 155, the regulation suggests that: "Possible attack impacts may include:
(a) Safe operation of vehicle affected;

(b) Vehicle functions stop working;
(c) Software modified, performance altered;
(d) Software altered but no operational effects;
(e) Data integrity breach;
(f) Data confidentiality breach;
(g) Loss of data availability;
(h) Other, including criminality."

However, from the author point of view, some of suggested possible impacts are vague or not straight forwards, such as the points of (b), (c), (e), (f), (g) and (h), so, this book suggests that the impact analysis should be done from the three points mentioned above and copied below, which fully cover the vehicle impacts:
- Safety
- Finance
- Privacy

3.4.7.2 *Cybersecurity Development Content*
The automotive ECU cybersecurity threats or attacks may apply the impacts by:
- executing a piece of code in the ECU to manipulate the vehicle, to do so, there are only two possible situations: either in the ECU production when the software is downloaded to the ECU, or during the updating software process where the new version software will be downloaded to the ECU.
- Accessing the ECU using the diagnostic services through the On-Board Diagnostic (OBD) connections or using the testing interfaces, such as JTAG, XCP, to modify the software contents or parameter values, or execute some service routines.
- modifying or interfering the communication with the ECU to manipulate the ECU's behavior.
- reversing engineering methods to retrieve the information from the ECU.

To most ECUs in Figure 3.4-32 Vehicle Network Structure, the first three threats in the list above are the development focus, the last one has very low attack feasibility according to the analysis based on the ISO 21434, and the application can do little about it, so it will not be discussed in this book.

So, to the automotive ECUs illustrated in Figure 3.4-32 Vehicle Network Structure except the Gateway and On-Board Diagnostic ECUs, the cybersecurity development contents are to prevent ECU from the first three attacks, for which, the measurements are to ensure:
- Trusted contents in the ECU
- Authenticated access to the ECU
- Authenticated communication with the ECU

The automotive ECU cybersecurity development is based on not only the attack impact, but also the risk analysis that is based on the attack effort and feasibility.
The cybersecurity threat risk analysis is guided by the ISO 21424, in which, the attack effort ratings are classified in Table 3.4-20 Attack Effort below:

Elapsed time		Specialist expertise		Knowledge of the item or component		Window of opportunity		Equipment	
Enumerate	Value	Enumerate	Value	Enumerate	Value	Enumerate	Value	Enumerate	Value
<= 1 day	0	Layman	0	Public	0	Unlimited	0	Standard	0
<= 1 week	1	Proficient	3	Restricted	3	Easy	1	Specialized	4
<= 1 month	4	Expert	6	Confidential	7	Moderate	4	Bespoke	7
<= 6 months	17	Multiple experts	8	Strictly confidential	11	Difficult/None	10	Multiple bespoke	9
> 6 months	19								

Table 3.4-20 Attack Effort

And the attack feasibility ratings are classified according to the attack effort:

Attack feasibility rating	Values
High	0 - 13
Medium	14 -19
Low	20 - 24
Very Low	>= 25

Table 3.4-21 Attack Feasibility

3.4.7.3 Cryptography

As mentioned in the section of 3.4.3 Electronic Architect Design, if the development has the cybersecurity requirements, then the microcontroller selected must have the built-in devices to provide the required the cryptographic algorithms and features, such as devices are called: Crypto Engine or Security Hardware Extension (SHE) or Security Peripheral in some cases. The commonly used cryptographic algorithms and features are listed in the Table 3.4-22 Common Cybersecurity Function below.

Cybersecurity Function Interface				
(Legend: App-> Application Mode Manager, Diag->Diagnostic Service, Seri->Serial Signal Manager, Cyber->Cybersecurity, Feat->Feature Function, AUTO-> AUTOSAR)				
Cybersecurity Function AEAD_Encrypt	Cyber	Diag Seri AUTO	Client / Server	Function: Authenticated encryption with associated data (AEAD) encryption Key Length in bits: 128, 256
Cybersecurity Function AEAD_Decrypt	Cyber	Diag Seri AUTO	Client / Server	Function: Authenticated encryption with associated data (AEAD) decryption Key Length in bits: 128. 256

Cybersecurity Function AES_ECB_Encrypt	Cyber	Diag Seri AUTO	Client / Server	Function: Electronic Code Book (ECB) AES encryption Key Length in bits: 128, 192, 256
Cybersecurity Function AES_ECB_Decrypt	Cyber	Diag Seri AUTO	Client / Server	Function: Electronic Code Book (ECB) AES decryption Key Length in bits: 128, 192, 256
Cybersecurity Function AES_CBC_Encrypt	Cyber	Diag Seri AUTO	Client / Server	Function: Cipher Block Chaining (CBC) AES encryption Key Length in bits: 128, 192, 256
Cybersecurity Function AES_CBC_Decrypt	Cyber	Diag Seri AUTO	Client / Server	Function: Cipher Block Chaining (CBC) AES decryption Key Length in bits: 128, 192, 256
Cybersecurity Function AES_CFB_Encrypt	Cyber	Diag Seri AUTO	Client / Server	Function: Cipher Feedback (CFB) AES encryption Key Length in bits: 128, 192, 256
Cybersecurity Function AES_CFB_Decrypt	Cyber	Diag Seri AUTO	Client / Server	Function: Cipher Feedback (CFB) AES decryption Key Length in bits: 128, 192, 256
Cybersecurity Function RSA_Encrypt	Cyber	Diag Seri AUTO	Client / Server	Function: Rivest–Shamir–Adleman (RSA) private - public key encryption Key Length in bits: 1024, 2048, 4096
Cybersecurity Function RSA_Decrypt	Cyber	Diag Seri AUTO	Client / Server	Function: Rivest–Shamir–Adleman (RSA) private - public key decryption Key Length in bits: 1024, 2048, 4096
Cybersecurity Function RSA_Signature_ Generate	Cyber	Diag Seri AUTO	Client / Server	Function: RSA signature generation Key Length in bits: 1024, 2048, 4096
Cybersecurity Function RSA_Signature_ Verify	Cyber	Diag Seri AUTO	Client / Server	Function: RSA signature verification Key Length in bits: 1024, 2048, 4096
Cybersecurity Function CMAC_Generate	Cyber	Diag Seri AUTO	Client / Server	Function: Cipher-based Message Authentication Code (CMAC) generation Key Length in bits: 128, 192, 256
Cybersecurity Function CMAC_Verifgy	Cyber	Diag Seri AUTO	Client / Server	Function: Cipher-based Message Authentication Code (CMAC) verification Key Length in bits: 128, 192, 256

Cybersecurity Function GMAC_Generate	Cyber	Diag Seri AUTO	Client / Server	Function: Galois Message Authentication Code (GMAC) generation Key Length in bits: 128, 192, 256
Cybersecurity Function CMAC_Verifgy	Cyber	Diag Seri AUTO	Client / Server	Function: Galois Message Authentication Code (GMAC) verification Key Length in bits: 128, 192, 256
Cybersecurity Function HASH_Generate	Cyber	Diag Seri AUTO	Client / Server	Function: Hashing Number generation Output Length in bits: 224, 256, 384, 512
Cybersecurity Function Random_Number	Cyber	Diag Seri AUTO	Client / Server	Function: Random Number generation
Cybersecurity Function Key_Derivation	Cyber	Diag Seri AUTO	Client / Server	Function: Key Derivation
Cybersecurity Function Key_Exchange	Cyber	Diag Seri AUTO	Client / Server	Function: Key exchange
Cybersecurity Function Certificate_Parse	Cyber	Diag Seri AUTO	Client / Server	Function: Certificate parse
Cybersecurity Function Certificate_Verify	Cyber	Diag Seri AUTO	Client / Server	Function: Certificate verify.

Table 3.4-22 Common Cybersecurity Function

The cybersecurity functions above are categorized as:
- Symmetric: Advanced Encryption Standard (AES), Data Encryption Standard (DES)

The symmetric encryption and decryption require that the information sender and receiver must use the same key to encrypt and decrypt the information transmitted. To do so, there must be the mechanisms to transmit and manage the secret keys, which, in turn, need the cybersecurity mechanisms for the protection, as well. The benefits of symmetric cryptographic algorithms are that they are fast and easy to implement.

The common symmetric algorithms are:
- o AES 128
- o AES 256
- o ECB
- o CBC
- o CMAC
- o GCM

3.4.7 Cybersecurity

In automotive ECU development, the symmetric cryptographic algorithms are often used in the serial communication authentication and secure boot due to the latency requirements, and they are used in the authenticated access protection because the key management can be shared to simplify the development, so that the symmetric cryptographic algorithms are the ones that are mainly implemented cybersecurity mechanisms in the ECUs.

- Asymmetric: Rivest-Shamir-Adleman (RSA) Cryptography, Elliptic Curve Cryptography (ECC).

Asymmetric cryptography uses pairs of keys: public keys and private keys, among which, the public keys may be known to others publicly, and the private keys are known only to the owner.

The asymmetric cryptographic algorithms are highly secure and easy to implement at the public key users' side, however, the disadvantages are that they are slow and more implementation effort at the private user's side.

In automotive ECUs, they are mainly used for the software production authentication and transfer the symmetric keys.

- SHA

The Secure Hashing Algorithm (SHA) is a mathematical function that calculates an arbitrary length data to generate the hash value:
  - For a specific data, the hash value is unique, that is: it is not possible for two different data to have the same hash value.
  - The hash value is invertible to the source data, that is impossible to reverse the source data from a given hash value

The hash algorithms can be used to protect the data's integrity, which is to detect if the data is modified by comparing the hash value against the re-calculated one.

The common SHA are: SHA-128, SHA-256, SHA-384, SHA-512.

- Random Number

The random number function is used to generate the unpredictable number for other cryptographical algorithms, such as key generation or initialization vector (IV) in encryption.

There are two commonly used mechanisms to generate the random numbers: hardware based True Random Number Generator (TRNG) and software-based Pseudo Random Number Generator (PRNG).

- Key management

To store the keys permanently that are used for the cryptographic algorithms and manage them including updating them, the microcontroller needs to have the special capacity and functional ability, for which, the essential rules are:
- The keys are device specific, so that the encryptions will be device specific to increase the security.
- The keys cannot be access by other application software, so that the possibility of attacking the keys can be avoided.
- Some of key are unknown to anyone in some cases, so that the keys are secured.

The microcontroller selection must pay attention to the built-in cybersecurity functionalities if the development is the cybersecurity relevant due to the reason that is described in the section of 3.4.4.1 Cybersecurity, because if the required cryptographic algorithms are not fully provided by the crypto engine in the microcontroller, then either the development is not feasible, or the development effort will be increased significantly.

3.4.7.4 Approach

The cybersecurity development approach for the required contents mentioned in the section of 3.4.7.2 Cybersecurity Development Content are following:

Trusted contents in the ECU

The measurements to provide the trusted contents in the ECU are divided into two parts:

- The first part is to ensure that only the authenticated contents will be programmed into the ECU.
- The second part is to make sure that the contents in the ECU are not modified by the unauthenticated activities.

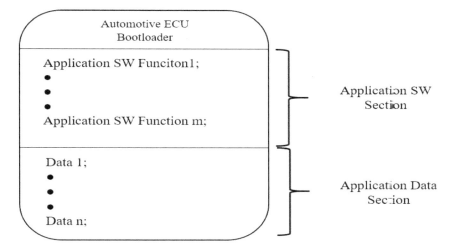

Figure 3.4-33 ECU NVM Structure

All contents in an automotive ECU are stored in the NVM which will be loaded into the CPU(s) to run once the ECU is powered on. The content structure is illustrated in Figure 3.4-33 ECU NVM Structure consisting three parts:

- Bootloader
- Application Software
- Application Data

The first part of approach to ensure an ECU security is to implement the automotive ECU bootloader that meets the cybersecurity requirements.

The automotive ECU bootloader is a software program that is used to program or update the contents in the ECUs, which makes use the services in the UDS and collaborates with the autosar NVM manager to erase and write the required contents in the ECUs.

The automotive ECU bootloader is always stored in the NVM, and loaded into the CPU only when it is needed.

3.4.7 Cybersecurity

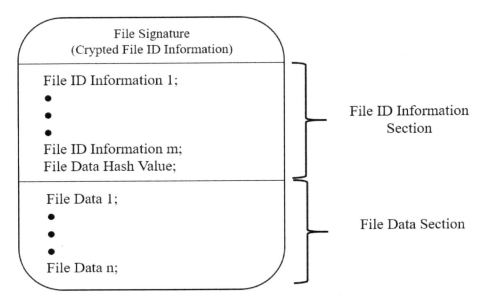

Figure 3.4-34 ECU Programming Structure

The contents to be programmed into an ECU have the structure illustrated in Figure 3.4-34 ECU Programming Structure, in which the File Signature is the crypted value of the File ID information including serial number, version number, target ECU ID, programming tool ID and other necessary identification information; the File Data Hash Value is the hashing result of File Data Section which may be either application software or application data, or the combination of both.

To ensure the ECU content security, the bootloader will follow the programming procedure illustrated in Figure 3.4-35 ECU Programming Procedure, in which:

- The programming activities must be done in the programming diagnostic session, which is controlled by using the diagnostic service of $10.
- Before the programming can be started, the authentication to program must be verified by using the Challenge-Response mechanism supported by the UDS 0x27 service (security access service) using either the public – private keys algorithm or the symmetry cryptography such as AES. (The detailed Challenge-Response mechanism is described and illustrated in Figure 3.4-36 ECU Unlock Procedure.)

3.4.7 Cybersecurity

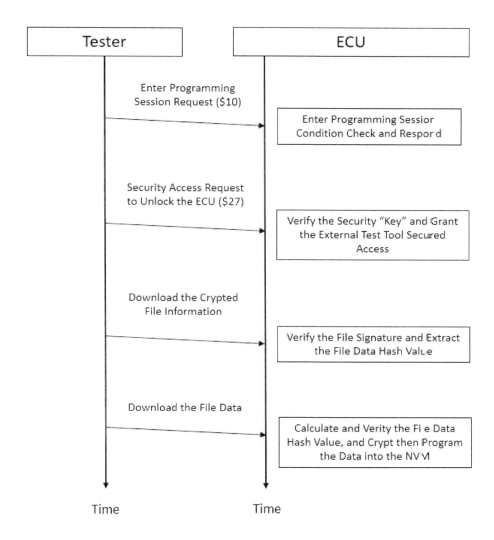

Figure 3.4-35 ECU Programming Procedure

- The bootloader must check the contents to be programmed against:
    - the authentication using the mechanisms of public – private keys and hash, which is to first calculate the hash value of File ID Information section contents, then encrypt the hash value using the software owner's private key as the signature that will be verified by the bootloader using the public key during the programming.
    - the integrity using the SHA algorithm. After the bootloader writes the application SW or application data into the NVM memory, it will calculate the hash value over the programmed memory, which will then be compared to the one in the File ID Information section. If the bootloader detects a mismatch between the two, then it will conclude that the content is invalid and therefore should not be used.

- If the SW and data contents are stored in the external NVM, then the contents need to be encrypted using symmetric keys to prevent the contents from the Differential Power Analysis (DPA) attacks. Encryption is the commonly used measure to protect the contents in the external NVM, because the external flash memory chip is connected to the microcontroller chip physically by soldering, so the NVM contents are vulnerable to the Differential Power Analysis (DPA) attacks, by which the attackers can un-solder the NVM chip from the original board and perform the DPA By analyzing the power consumption or radiation from the NVM chip, so that the attackers can recover or manipulate the chip contents.

The second part of approach to ensure an ECU security is to use the secure boot procedure, which is invoked at every time when the ECU has the power on reset consisting of the following steps:

- First, the built-in Security Hardware Extension illustrated in Figure 2.2-5 Security Hardware Extension in the microcontroller will verify the integrity and authentication of the built-in firmware that is encrypted and stored in its EEPROM to make sure that those contents are trusted by decrypting them using the designated symmetry key. This part is done by the microcontroller hardware.
- Second, after the built-in Security Hardware Extension has the firmware running, it will load and verify the integrity and authentication of the application boot loading software into the master boot CPU that could be one of cores in Figure 2.2-4 Microcontroller by decrypting the contents using the specific designated symmetry key. Then the built-in Security Hardware Extension will transfer the execution control to the application boot loading software.
- Then the application boot loading software will load and verify the integrity and authentication of rest application software and data by decrypting the contents using the specific designated symmetry keys.

Authenticated access to the ECU

All the following accesses must be secured using the locking and unlocking mechanisms illustrated in Figure 3.4-36 ECU Unlock Procedure to make sure that only the authenticated persons can access the ECUs if the access is related the cybersecurity, the accesses consist of:

- All the access to the ECU development interfaces, such as JTAG, XCP,
- Read the confidential information using the diagnostic services like the $22, such as reading of ECU ID, critical product parameters, etc.
- Modify the critical parameters using the diagnostic service $2E.
- Request to run certain diagnostic routines using the $2F or $31, such as erasing memory, executing safety or security related routines.

The default state of the ECU development interfaces is locked, and the ECU must lock the interfaces all the time after the production except the time when the development is ongoing or the quality return analysis is ongoing, which is done by the security mechanisms implemented by the microcontroller hardware to check and lock the interfaces at every time when the ECU has the power on reset even before any software or firmware is able to run, in this way, the interfaces are secured.

For the diagnostic services requests to the ECU via either the OBD connector or remote diagnostic gateway ECU, it will use the Challenge-Response mechanism provided by the diagnostic service $27.

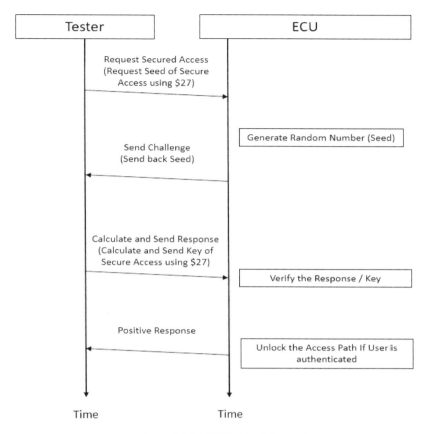

Figure 3.4-36 ECU Unlock Procedure

Authenticated communication with the ECU

For the ECUs inside of vehicle network, the communication between them is strictly configured and management, the characteristic of such communication is: which messages that an ECU should receive and send out are configured during the development, and will be checked during the runtime, so they are secured in normal cases. However, to prevent the communication from the high-tech attacks, such as inserting some messages into the communication, or interfering the communication by replaying the valid messages, some security measures should be implemented.

- Message authentication or encryption

Message authentication: if the protected messages just need the data authenticity (integrity), and the message data are allowed to be seen by others, then the data authenticity implementation for the protected messages will be enough, such as CMAC, GMAC.

Message encryption: if the protected messages need not only the data authenticity (integrity) but also the data confidentiality, the authenticated encryption algorithm such as GCM should be used to cipher the messages.

For both scenarios above, the symmetric key algorithms will be used to increase the performance. And the message senders will encrypt the protected messages using the dedicated keys, and the message receivers will decrypt the message using the same key.

- Message Counter

Message counter is used to prevent the messages from replaying by attackers, which is that the sender and receiver have the same counter for the protected messages, and the sender will embed the current counter value into the message protect values, then send out the values along with the messages; the receiver should count its own counter value that is used to against the embedded counter value to check if the messages order are expected.

To use the message counter mechanism, the counter synchronization mechanism should be established between the senders and the receivers at the time of the initialization from the reset and from the fault recovery.

3.4.7.5   UN ECE 155 / 156 compliance

The UN ECE 155 / 156 regulations below are mandatory for the vehicles to be sold in the Europe, North America, Japan and Korea. So, this section will provide full implementation to cover the technical requirements from the regulations.

- UN ECE 155: UN Regulation on uniform provisions concerning the approval of vehicles with regard to cyber security and of their cybersecurity management systems
- UN ECE 156: UN Regulation on uniform provisions concerning the approval of vehicles with regards to software update and software updates management system

The UN ECE 156 is about the vehicle software updates and update management system, which does not mainly concern the ECU development, rather, it is mainly about the ECU development organization processes and the updating procedures, so it is out of the scope of the book.

However, to facilitate the UN ECE 156 compliance, here are the brief descriptions and suggestions about it: all the detailed requirements in this regulation are documented in the chapter 7: "General specifications", which covers following:

- Update content identification, which requires the vehicle manufacturer to identify and record what the update contents are, what are the target vehicles. To satisfy it, the vehicle manufacturer needs to have the information management system, which should enable to retrieve the identification information both outside and inside of the target vehicle.
- Update security and integrity, which requires the vehicle manufacturer to ensure that the update contents are securely transferred to the target vehicles and ensure the security after transfer. To satisfy it, the vehicle manufacturer needs to implement the security measures at three points:
  - o At the manufacturing site, which needs the secure information management system.
  - o During the transmission between the manufacturing site and the target vehicle, which needs the secure communication channel.
  - o At the target vehicle, which can be done by implementing the measures in 3.4.7.4 Approach.

- Update impact, which requires that vehicle manufacturer should identify and evidence what effects the update will have
  - To its own functionalities, to satisfy it, for the ECU developers, the change list and the verification reports are needed.
  - To other systems in the vehicle, to satisfy it, the vehicle manufacturer needs to have
    - the change list that covers the update contents
    - the verification reports that evidence the update functionalities.
  - To the vehicle users, to satisfy it, the vehicle manufacturer needs to:
    - Ensure the vehicle safety during and after the updating.
    - Inform the vehicle users about the updating.

The UN ECE 155 is the regulation about vehicle cybersecurity and cybersecurity management system, among which, the detailed cybersecurity technical requirements are specified in the Annex 5: "List of threats and corresponding mitigations". Which consists of three parts:

- Part A of the annex: "Vulnerability or attack method related to the threats" describes the baseline for threats, vulnerabilities and attack methods.
- Part B of the annex: "Mitigations to the threats intended for vehicles" describes mitigations to the threats which are intended for vehicle types.
- Part C of the annex: "Mitigations to the threats outside of vehicles" describes mitigations to the threats which are intended for areas outside of vehicles, e.g., on IT backends, which are the focus of the book, so, for which, only the brief suggestions will be provided.

This book will provide the solutions to prevent the ECUs from the threats, vulnerabilities and attack methods in the Part A, and focus on the details which will fully cover the mitigations in the Part B against the threats for the vehicle.

Note: In the following items, all requirements from the Part A of the annex are copied here starting with bold font and with the original section numbers, for example, the "4.3.1" in the items below, is from the Part A of annex 5, not the section number from this book. For each section of requirements, the protection measures will be provided in the section that is started with the bold font: "**Protection**".

4.3.1 Threats regarding back-end servers related to vehicles in the field
1 Back-end servers used as a means to attack a vehicle or extract data
    1.1 Abuse of privileges by staff (insider attack)
    1.2 Unauthorized internet access to the server (enabled for example by backdoors, unpatched system software vulnerabilities, SQL attacks or other means)
    1.3 Unauthorized physical access to the server (conducted by for example USB sticks or other media connecting to the server)
2 Services from back-end server being disrupted, affecting the operation of a vehicle
    2.1 Attack on back-end server stops it functioning, for example t prevents it from interacting with vehicles and providing services they rely on
3 Vehicle related data held on back-end servers being lost or compromised ("data breach")
    3.1 Abuse of privileges by staff (insider attack)
    3.2 Loss of information in the cloud. Sensitive data may be lost due to attacks or

accidents when data is stored by third-party cloud service providers

3.3 Unauthorized internet access to the server (enabled for example by backdoors, unpatched system software vulnerabilities, SQL attacks or other means)

3.4 Unauthorized physical access to the server (conducted for example by USB sticks or other media connecting to the server)

3.5 Information breach by unintended sharing of data (e.g., admin errors)

Protection: The requirements above are for the back-end server, not for the ECUs, they are out of the book's focus, but some of them will impact the ECUs functionalities, so the suggestions are:

- For the first part: "Back-end servers used as a means to attack a vehicle or extract data" including the three subsections, the organization needs to establish the processes about the back-end server management and maintenance to prevent those issues which are commonly known in IT industry.
- For the second part regarding to the server operation issue, and the third part regarding to the data being lost or compromised:
  - At the server-end, the "fail-safe" mechanisms are needed, which are commonly implemented in all the IT servers.
  - At the vehicle-end, the relative "fail-safe" should be considered to prevent the ECUs from not only the attacks, but also normal interruptions, and the protection mechanisms against data lost and operation interruptions are commonly implemented in the automotive ECUs.

4.3.2 Threats to vehicles regarding their communication channels

4 Spoofing of messages or data received by the vehicle

4.1 Spoofing of messages by impersonation (e.g., 802.11p V2X during platooning, GNSS messages, etc.)

4.2 Sybil attack (in order to spoof other vehicles as if there are many vehicles on the road)

Protection: According to the network characteristics in a vehicle illustrated in Figure 3.4-32 Vehicle Network Structure that is described at the beginning of this section, for the ECUs that are behand the gateway ECU, those kinds of messages are not applicable, and for attacks from outside of the vehicle network, they will be filtered by the Gateway ECU.

The cybersecurity issues are applicable to the ECUs that are located before the gateway ECU, they are connected to the outsides directly, which is out of this book scope.

5 Communication channels used to conduct unauthorized manipulation, deletion or other amendments to vehicle held code/data

5.1 Communications channels permit code injection, for example tampered software binary might be injected into the communication stream

5.2 Communications channels permit manipulate of vehicle held data/code

5.3 Communications channels permit overwrite of vehicle held data/code

5.4 Communications channels permit erasure of vehicle held data/code

5.5 Communications channels permit introduction of data/code to the vehicle (write data code)

Protection: For the ECUs inside of the vehicle network, there are only three ways to modify either the software code or the data in the ECUs via the communication channels:

- Development Interfaces
- Diagnostic services, such as $2E, $2F or $31

- Programming or updating the ECU contents via the bootloader

To prevent the ECUs from the threats coming from the development interfaces or from the diagnostic services, the ECUs need to implement the measures of "Authenticated access to the ECU" in 3.4.7.4 Approach.

To prevent the ECUS from the threats coming from the bootloader, the first part of mechanism mentioned in "Trusted contents in the ECU" in 3.4.7.4 Approach needs to be implemented.

6 Communication channels permit untrusted/unreliable messages to be accepted or are vulnerable to session hijacking/replay attacks

6.1 Accepting information from an unreliable or untrusted source

6.2 Man in the middle attack/ session hijacking

6.3 Replay attack, for example an attack against a communication gateway allows the attacker to downgrade software of an ECU or firmware of the gateway

Protection: For the ECUs inside of the vehicle network, the "Man in the middle attack/ session hijacking" is not applicable; and all received messages are from configured senders that are trusted. The Gateway ECU security functions are out of the scope of this book.

However, to prevent the communications from the high-tech attacks and the replay attacks, the measures of "Authenticated communication with the ECU" in 3.4.7.4 Approach need to be implemented.

7 Information can be readily disclosed. For example, through eavesdropping on communications or through allowing unauthorized access to sensitive files or folders

7.1 Interception of information / interfering radiations / monitoring Communications

7.2 Gaining unauthorized access to files or data

Protection: For the ECUs inside of the vehicle network, to prevent the ECUs from the "Interception of information / interfering radiations / monitoring Communications", the measures of "Authenticated communication with the ECU" in 3.4.7.4 Approach need to be implemented, to prevent the ECUs from the "Gaining unauthorized access to files or data", the measures of "Authenticated access to the ECU" in 3.4.7.4 Approach need to be implemented.

8 Denial of service attacks via communication channels to disrupt vehicle functions

8.1 Sending a large number of garbage data to vehicle information system, so that it is unable to provide services in the normal manner

8.2 Black hole attack, in order to disrupt communication between vehicles the attacker is able to block messages between the vehicles

Protection: Inside of the vehicle network, all messages are configured based on the design, so it is very unlikely for any ECU to send out any un-configured message.

If the attacks are from the high-tech message's insertions, then they will be low feasibility according to the ISO 21434;

If the attacks are from outside of the vehicle network, the Gateway ECU will filter the communication traffic and protect the ECUs.

9 An unprivileged user is able to gain privileged access to vehicle systems

9.1 An unprivileged user is able to gain privileged access, for example root access

Protection: The access should be managed by the organization processes, which should follow the requirements of UN ECE 155 / 156, especially the ones regarding to the management processes, which is not a development topic and out of this book scope.

10 Viruses embedded in communication media are able to infect vehicle systems

     10.1 Virus embedded in communication media infects vehicle systems

Protection: For the ECUs inside of the vehicle network, the only possibility for viruses to act in an ECU is to introduce the viruses via the bootloader, and to prevent the ECUs from the threats coming from the bootloader, the mechanisms mentioned in "Trusted contents in the ECU" in 3.4.7.4 Approach need to be implemented.

11 Messages received by the vehicle (for example X2V or diagnostic messages), or transmitted within it, contain malicious content

     11.1 Malicious internal (e.g., CAN) messages

     11.2 Malicious V2X messages, e.g., infrastructure to vehicle or vehicle-vehicle messages (e.g., CAM, DENM)

     11.3 Malicious diagnostic messages

     11.4 Malicious proprietary messages (e.g., those normally sent from OEM or component/system/function supplier)

Protection: Inside of the vehicle network, all the accesses to the test or debug interfaces of ECUs are protected using the approach of "Authenticated Access to the ECU" in 3.4.7.4 Approach that requires the Challenge-Response verification for the accesses; and all the messages that will change or modify any information inside of ECUs must use the diagnostic services that must pass the Challenge-Response and the RSA authentication mechanisms implemented using the approach in "Trusted contents in the ECU" in 3.4.7.4 Approach before being able to update or modify any data or code.

4.3.3. Threats to vehicles regarding their update procedures

12 Misuse or compromise of update procedures

     12.1 Compromise of over the air software update procedures. This includes fabricating the system update program or firmware

     12.2 Compromise of local/physical software update procedures. This includes fabricating the system update program or firmware

     12.3 The software is manipulated before the update process (and is therefore corrupted), although the update process is intact

     12.4 Compromise of cryptographic keys of the software provider to allow invalid Update

Protection: The measures against the threats above are described in the approach of "Trusted contents in the ECU" in 3.4.7.4 Approach, in which the update procedures will be verified against the authentications, the update contents will be verified against the "signature" using the public – private key mechanism. The cryptographic keys used above are managed by the built-in crypto engine in the microcontrollers, which will prevent the keys from access by anyone, even by the developers.

13 It is possible to deny legitimate updates

     13.1 Denial of Service attack against update server or network to prevent rollout of critical software updates and/or unlock of customer specific features

Protection: There are only three locations where such attacks may happen:

- The Server end

- The ECU end
- The communication path between the server and the ECU

It is unlikely happen at the ECU end if the bootloader measures described in the "Trusted contents in the ECU" in 3.4.7.4 Approach are implemented.

The attacks at both the server-end and the communication path are out of the scope of this book, and currently there are the commonly available measures at the server-end to protect the IT server, and there are the commonly available measures about the telecommunication and the network to protect the communication path, so they are low feasibility and low impact according to the ISO 21434.

Note: there is not the number 14 item in the regulation original document.

4.3.4 Threats to vehicles regarding unintended human actions facilitating a cyber attack

15 Legitimate actors are able to take actions that would unwittingly facilitate a cyber- attack

    15.1 Innocent victim (e.g., owner, operator or maintenance engineer) being tricked into taking an action to unintentionally load malware or enable an attack

    15.2 Defined security procedures are not followed

Protection: Those kinds of attacks are purely management topics, which are out of the scope of the book.

4.3.5 Threats to vehicles regarding their external connectivity and connections

16 Manipulation of the connectivity of vehicle functions enables a cyber- attack, this can include telematics; systems that permit remote operations; and systems using short range wireless communications

    16.1 Manipulation of functions designed to remotely operate systems, such as remote key, immobilizer, and charging pile

    16.2 Manipulation of vehicle telematics (e.g., manipulate temperature measurement of sensitive goods, remotely unlock cargo doors)

    16.3 Interference with short range wireless systems or sensors

Protection: All the attacks from outside of the vehicle network will be protected by the Gateway ECU that is out of the scope of this book. The remote actions, such as to use the key forbs to operate the doors, windows, liftgate, root, will be verified by either the key forb ECU or the Gateway ECU using the Challenge-Response mechanism.

17 Hosted 3rd party software, e.g., entertainment applications, used as a means to attack vehicle systems

    17.1 Corrupted applications, or those with poor software security, used as a method to attack vehicle systems

Protection: In order for an automotive ECU as a means to attack vehicle system, it has to send out the information using the communication channels that it has, however, according to the network characteristics in a vehicle illustrated in Figure 3.4-32 Vehicle Network Structure that is described at the beginning of this section, all messages are configured based on the design, so it is very unlikely for any ECU to send out any un-configured message.

And for any ECU software including the third-party software to be effective in a vehicle system, first the software has to be programmed using the bootloader into an ECU, and the programming contents must be signed by the software owners, so such threats will unlikely happen.

For the attacks via the communication messages from outside of the vehicle network, they will be filtered by the Gateway ECU. The cybersecurity issues are applicable to the ECUs like the CarPlay ECU, GPS ECU, that are located before the gateway ECU, they are connected to the outsides directly, which is out of this book scope.

18 Devices connected to external interfaces e.g., USB ports, OBD port, used as a means to attack vehicle systems

18.1 External interfaces such as USB or other ports used as a point of attack, for example through code injection

18.2 Media infected with a virus connected to a vehicle system

18.3 Diagnostic access (e.g., dongles in OBD port) used to facilitate an attack, e.g., manipulate vehicle parameters (directly or indirectly)

Protection: For the ECUs inside of the vehicle network, the measures of "Authenticated access to the ECU" in 3.4.7.4 Approach need to be implemented to prevent the ECUs from the threats coming from the external interfaces, such as the XCP, JTAG and OBD. Generally, those ECUs do not have the USB ports.

4.3.6 Threats to vehicle data/code

19 Extraction of vehicle data/code

19.1 Extraction of copyright or proprietary software from vehicle systems (product piracy)

19.2 Unauthorized access to the owner's privacy information such as personal identity, payment account information, address book information, location information, vehicle's electronic ID, etc.

19.3 Extraction of cryptographic keys

Protection:

- Regarding to the treats in 19.1 above, the ECU SW and data contents are encrypted using the implemented measures in the "Trusted contents in the ECU" in 3.4.7.4 Approach if they are stored in the external NVM, and regarding to the threats of reading software contents, neither the ECU bootloader nor the diagnostic services can upload the ECU software contents out of the ECU, so the contents are safe.
- Regarding to the threats in 19.2, the measures of "Authenticated access to the ECU" in 3.4.7.4 Approach need to be implemented.
- Regarding to the threats in 19.3, the cryptographic keys are always stored in the built-in crypto engine of the microcontroller which cannot be extracted by anyone, even the developers.

20 Manipulation of vehicle data/code

20.1 Illegal/unauthorized changes to vehicle's electronic ID

20.2 Identity fraud. For example, if a user wants to display another identity when communicating with toll systems, manufacturer backend

20.3 Action to circumvent monitoring systems (e.g., hacking/ tampering/ blocking of messages such as ODR Tracker data, or number of runs)

20.4 Data manipulation to falsify vehicle's driving data (e.g., mileage, driving speed, driving directions, etc.)

20.5 Unauthorized changes to system diagnostic data

Protection:

- To avoid the threats from the issues in 20.1, the access to the secret information in the ECUs, such as the vehicle's electronic ID, mileage, should be verified using the Challenge-Response mechanism in the "Authenticated access to the ECU" in 3.4.7.4 Approach.

- Regarding to the threats in 20.2, if fraud messages are from the ECUs inside of the vehicle network, then the measures of "Authenticated access to the ECU" in 3.4.7.4 Approach need to be implemented to prevent the ECUs from any unintended modification to send the fraud messages; if the fraud messages are sent out from the ECUs outside of the vehicle network, then it is out of this book scope, and the suggestions are to design the "hand-shaking" or authentication signals for such communication such as the measures of "Authenticated access to the ECU" in 3.4.7.4 Approach.

- Regarding to the threats in 20.3, if the threats are from the ECUs inside of the vehicle network, then the measures of "Authenticated access to the ECU" in 3.4.7.4 Approach need to be implemented to prevent the ECUs from any unintended modification to have such actions; if the action messages are sent out from the ECUs outside of the vehicle network, then it is out of this book scope, and the suggestions are to design the "hand-shaking" or authentication signals for such communication such as the measures of "Authenticated access to the ECU" in 3.4.7.4 Approach.

- Regarding to the data manipulation in 20.4, if the data manipulation are targeted at the ECUs inside of the vehicle network, then the measures of 'Authenticated access to the ECU" in 3.4.7.4 Approach need to be implemented to prevent the ECUs from any unintended data modification; if the data manipulation happens outside of the vehicle network, then it is out of this book scope, and the suggestions are to design the "hand-shaking" or authentication signals for such communication such as the measures of "Authenticated access to the ECU" in 3.4.7.4 Approach, or the measures of "Authenticated communication with the ECU" in 3.4.7.4 Approach need to be implemented to enhance the communication confidentiality and the integrity.

- Regarding to the unauthorized changes to system diagnostic data in 20.5, the measures of "Authenticated access to the ECU" in 3.4.7.4 Approach need to be implemented to prevent the ECUs from any unintended data modification.

21 Erasure of data/code

21.1 Unauthorized deletion/manipulation of system event logs

Protection: For the ECUs inside of the vehicle network, to prevent them from the unauthorized deletion or manipulation of system event logs, the measures of "Authenticated access to the ECU" in 3.4.7.4 Approach need to be implemented, because the development interfaces and the diagnostic routine execution service request ($31) are the only way to change the system vent logs.

22 Introduction of malware

22.1 Introduce malicious software or malicious software activity

Protection: For the ECUs inside of the vehicle network, to prevent them from the introducing malicious software, the measures of "Trusted contents in the ECU" in 3.4.7.4 Approach need to be implemented.

23 Introduction of new software or overwrite existing software

23.1 Fabrication of software of the vehicle control system or information system
Protection: The updates and modifications to the ECU software will be verified using the measures of "Trusted contents in the ECU" in 3.4.7.4 Approach, in which, the bootloader will ensure the validity of programmed or updated software to the target ECUs.

24 Disruption of systems or operations
24.1 Denial of service, for example this may be triggered on the internal network by flooding a CAN bus, or by provoking faults on an ECU via a high rate of messaging
Protection: For the ECUs inside of the vehicle network, the communication messages in the vehicle network are configured according to the design; if the attacks are from the high-tech message insertions, then they are low feasibility and low impact according to the ISO 21434.

25 Manipulation of vehicle parameters
25.1 Unauthorized access of falsify the configuration parameters of vehicle's key functions, such as brake data, airbag deployed threshold, etc.
25.2 Unauthorized access of falsify the charging parameters, such as charging voltage, charging power, battery temperature, etc.
Protection: The "Authenticated access to the ECU" will protect the ECU from every access of falsify which will be either using the debug interfaces, such XCP, JTAG, or using the diagnostic services, and both of them will have to go through the Challenge-Response mechanism if the measures in "Authenticated access to the ECU" in 3.4.7.4 Approach are implemented.

4.3.7 Potential vulnerabilities that could be exploited if not sufficiently protected or hardened

26 Cryptographic technologies can be compromised or are insufficiently applied
26.1 Combination of short encryption keys and long period of validity enables attacker to break encryption
26.2 Insufficient use of cryptographic algorithms to protect sensitive systems
26.3 Using already or soon to be deprecated cryptographic algorithms
Protection: For the ECUs inside of the vehicle network, the cryptographic algorithms mentioned in the section of 3.4.7.3 Cryptography are reasonably enough, which are compliant with the current relative industrial standards. The applications and implementations of the cryptographic algorithms in 3.4.7.4 Approach cover all aspects reasonably based on the impact analysis according to the ISO 21434.

27 Parts or supplies could be compromised to permit vehicles to be attacked
27.1 Hardware or software, engineered to enable an attack or fails to meet design criteria to stop an attack
Protection: This kind of issues belong to the development management, not the technical development topics. However, If the ECU developers implement the measures mentioned in 3.4.7.4 Approach for the ECUs inside of the vehicle network, and follow the ASPICE to execute the development activities especially the verifications, then the protections are reasonably enough based on the impact analysis according to the ISO 21434.

28 Software or hardware development permits vulnerabilities

28.1 Software bugs. The presence of software bugs can be a basis for potential exploitable vulnerabilities. This is particularly true if software has not been tested to verify that known bad code/bugs is not present and reduce the risk of unknown bad code/bugs being present

28.2 Using remainders from development (e.g., debug ports, JTAG ports, microprocessors, development certificates, developer passwords, …) can permit access to ECUs or permit attackers to gain higher privileges

Protection:

Regarding to the threats in 28.1, the weak development quality may lead to some vulnerabilities to be attacked, which is out of this book's scope, however, the suggestions are to implement the quality control processes by following the ASPICE, and for the ECUs inside of the vehicle network, the measures mentioned in the section of 3.4.7.4 Approach are reasonably enough.

Regarding to the threats in 28.2, the suggestions are to implement the management processes to cover those aspects.

29 Network design introduces vulnerabilities

29.1 Superfluous internet ports left open, providing access to network systems

29.1 Circumvent network separation to gain control. Specific example is the use of unprotected gateways, or access points (such as truck-trailer gateways), to circumvent protections and gain access to other network segments to perform malicious acts, such as sending arbitrary CAN bus messages

Protection: For the ECUs inside of the vehicle network, the measures mentioned in the section of 3.4.7.4 Approach above are reasonably enough. However, the ECUs, such as the Gateway ECU, GPS ECU, CarPlay ECU, need more comprehensive cybersecurity protections, which is out of this book's scope.

Note: there is not the number 30 item in the regulation original document.

31 Unintended transfers of data can occur

31.1 Information breach. Personal data may be leaked when the car changes user (e.g., is sold or is used as hire vehicle with new hirers)

Protection: The protections for those vulnerabilities belong to the system design in the ECUs, such as the Gateway ECU, GPS ECU, CarPlay ECU, that needs more comprehensive cybersecurity measures, which is out of this book's scope. However, the suggestions are to implement the user identification measures when the personal data are used, such as the biological identification using facial recognition, fingerprint, personal passcode, etc., which are commonly implemented in the smart phone, tablet computers.

32 Physical manipulation of systems can enable an attack

32.1 Manipulation of electronic hardware, e.g., unauthorized electronic hardware added to a vehicle to enable "man-in-the-middle" attack Replacement of authorized electronic hardware (e.g., sensors) with unauthorized electronic hardware Manipulation of the information collected by a sensor (for example, using a magnet to tamper with the Hall effect sensor connected to the gearbox)

Protection: The attacks to those vulnerabilities are out of reasonably consideration to the automotive ECUs, which are low feasibility according to the ISO 21434.

Check List 9 - Cybersecurity

- The built-in Security Hardware Extension (SHE)
    - Do the symmetric cryptographic algorithms (AES, DES, CBC, ECB, CFB, CMAC, GMAC) meet the system requirement?
    - Do the asymmetric cryptographic algorithms (RSA, ECC) meet the system requirement?
    - Do the Secure Hashing Algorithm (SHA) meet the system requirement?
    - Does the Random Number function (TRNG, PRNG) meet the system requirement?
    - Does the key management (capacity, key update methods, secret counter) meet the system requirement?
    - If there is any cryptographic algorithm that the SHE cannot provide, then can it be implemented based on the isolation between the "Secure World" and the "Non-Secure World"?
- Implementation of trusted contents in the ECU
    - Does the bootloader implement the Challenge-Response mechanism to unlock the ECU before the programming is started?
    - Does the bootloader implement the content authentication (RSA) and the content integrity (SHA) checking mechanism?
    - Is the content authentication or content confidentiality implemented if the contents are stored in the NVM?
    - Is the secure boot procedure implemented?
    - Are the development interfaces (JTAG, XCP) locked by the microcontroller hardware settings at the beginning of secure boot?
- Implementation of trusted access to the ECU
    - Is all the access to the ECU development interfaces, such as JTAG, XCP, protected by the Challenge-Response mechanism?
    - Is all the reading of confidential information, such as ECU ID, critical product parameters, using the diagnostic services like the $22 protected by the Challenge-Response mechanism?
    - Are all the modifying critical parameters using the diagnostic service $2E protected by the Challenge-Response mechanism?
    - Is all the request to run critical diagnostic routines using the $2F or $31, such as erasing memory, executing safety or security related routines, protected by the Challenge-Response mechanism?
- Implementation of authenticated communication with the ECU
    - Are the critical message authentication mechanisms implemented?
    - Are the message integrity mechanisms implemented?
    - Are the critical message anti-replay mechanisms implemented?

○

3.5 System Verification

The automotive ECU system verification is divided into two types of verification:

- System Integration Verification
- System Black Box verification

All functionality tests that verify the system behaviors' correctness should be done by the system integration test, the reasons for which are:

- In the integration tests, not only the external behaviors of the ECU under tests can be observed, but also the internal data values and status that derive the external behaviors can be observed as well, based on both of which, the judgements can be made to determine if the external behaviors are correct.
- In the integration tests, the testing environment can be manipulated to simulate the input signals to the integrated component under test to trigger all the scenarios, so the tests can be accurate and complete.

The system black box verification that is sometimes called the system test is only about two types of tests: Environment test and time test, in which, the system under test operates in the real or simulated the operation environment to test the system performance.

The most important system verification strategy is the regression verification strategy, because almost all the automotive ECU projects are the carried over projects, or the application projects that are developed based on the existed products, and it is hardly to get the awarded projects from the OEMs without the existed products.

All the verifications in the situations above are the regression verifications

As the connected vehicle technology progresses, especially the software update over the air is more commonly used in the automotive industry, the regression verification will directly impact the development efficiency and performance.

And in the UN ECE 156: "UN Regulation on uniform provisions concerning the approval of vehicles with regard to cyber security and of their cybersecurity management systems", which emphases that the vehicle manufacturer should verify the software impact from the updates, both to the internal and external functionalities in the vehicle, which makes the regression verification even more critical.

Based on the Data Drive approach, this book provides the clear and comprehensive regression verification strategy, which reveals clearly what impact the updates will have based on the data flow in the system operation concept.

The most efficient system verification way is the simulation test:

- The verification can be done even when the needed hardware and software platforms are not ready.
- The verification can be done at the early phase of the development, especially the system operation concept can be only verified by the simulation test, which will reveal the potential issues at the early time to make the development more efficient.
- The verification can be done more thoroughly and accurately because the execution environments that are simulated can be manipulated using the designed data or scenarios, some of which are impossible to be tested in the real execution environment.

The prerequisites of system verification are that:
- The system requirement and design specifications are ready and clearly understood.
- The verification strategy is defined including
  - the scope: what is the device under test (DUT)
  - the approach: what is the method to verify the DUT: simulation or in real execution environment; Hardware in Loop (HIP) or Software in Loop (SIP).

The system verification is to test the implementations against the requirements and design specifications, and the specifications are specified either using the text tools, such as IBM DOORS or PTC Integrity, or the notation tools, such as the SysML that includes 9 types of diagrams. The development of those specifications is based on the developers' experience, so, the potential issues of which are that there is neither the clearly defined explicit and complete approach to design the specifications, nor is there the clearly defined explicit and complete method to fully cover all the system, which will cause issues in the integration and verification.

For the text specified specifications, the issues will include that the text specifications are prone to ambiguous and incomplete, and it is difficult to figure out the logic relationships in the specifications, whose consequence is that the specifications may be inconsistent, incomplete and inaccurate, then it will cause the issues in the integration and verification.

For the notation specified specifications, the issues include that it is difficult to fully specify the system functionalities, and it is difficult to use the notations in the entire development team, and it is difficult to figure out the relationship in all the diagrams used in the development.

The common issues in the system integration are:
- The component interfaces do not match,
- The required operation platforms do not match including the partitions, cores.
- The required time conflicts each other, either the running sequences do not match, or the scheduled time is not correct.

This book provides the clearly defined and comprehensive system integration and verification approach based on the system operation concept described in 3.4.1 System Operation Concept Design Comparing with the conventional integration above, the Data Driven approach has the following advantages:
- The data types and their interactions are clearly defined by the system operation concept.

To each system output data, the system operation concept clearly defines all the needed data including the data type, resolution and characteristics and the operations between those data. The system operation concept is the system development foundation, which should be verified at the beginning of the development. So, the definitions in it must be correct and clear.
- Integration paths are clearly defined by the system operation concept.

Every data in the concept is an integration element, and data formulas are the integration paths.

Every data in the concept is an integration test case, and data formulas are the integration test execution sequence.

Since the system operation concept defines the data relationships, then the integration procedures can be done by following the relationships. Meanwhile, according to the relationships, the integration priorities, such as safety and security relevant elements, can be identified.

- Regression paths are clearly defined by the system operation concept.

Every data and formulas in the system operation concept will be relevant and impact the output signals, which defines the regression test scope:

- if a data is changed, then the result data that is derived using the data will be impacted, in which, the data change includes both data value change and data timing change, which can be caused by either transformation or transmission.
- if a formula is changed, then the result data that is derived using the formula will be impacted, in which, the formula change can be caused by either software code change or software execution resource change, such as the processor, RAM, ROM.

3.5.1 Test Case Setup

3.5.1.1 *Minimum Test Cases Number Calculation*

The test case granularity is one of key factors about the testing quality, the minimum test cases should be at least achieved for all the quality levels. And furthermore, the components with the different critical characteristics should have different granularity test, such as the components with different ASIL levels should have different granularity test, i.e., the ASIL D component should be tested in more detail than the one with ASIL C; the same granularity requirement should be applied for the cybersecurity feature, as well.

One of example is Ford ECU Software Testing Requirements, in which, Ford specifies software test granularity requirements based on the Functional Importance Classifications: Class C (most important), Class B (middle important), Class A (less important), for which, the Class C software must have the required highest test granularity.

There is not the way to decide the maximum test case number for a component, but there is the way to calculate the minimum test cases number in the following situations if the operation logics in the device under test (DUT) are expressed as below:

Logic Expression

To test the logic expression: x1 & x2, ..., xn, the minimum test cases should be n+1 cases, which should test each x1, ... xn as false plus one test case that test all the elements are true.

To test the logic expression: x1 || x2, ..., xn, the minimum test cases should be n+1 cases, which should test each x1, ... xn as true plus one test case that test all the elements are false.

Parameter Range

To test the open parameter range: (x, y), that is: x < Test Value < y, the minimum test cases should be 3 cases, which should be:

- Test Case 1: Test Value < x;
- Test Case 2: x < Test Value < y;
- Test Case 3: Test Value > y.

To test the one close end parameter range: [x, y), that is: x <= Test Value < y, the minimum test cases should be 4 cases, which should be:

- Test Case 1: Test Value < x;

- Test Case 2: Test Value = x;
- Test Case 3: x < Test Value < y;
- Test Case 4: Test Value > y.

To test the two close end parameter range: [x, y], that is: x <= Test Value <= y, the minimum test cases should be 5 cases, which should be:

- Test Case 1: Test Value < x;
- Test Case 2: Test Value = x;
- Test Case 3: x < Test Value < y;
- Test Case 4: Test Value = y.
- Test Case 4: Test Value > y.

Mathematical Expression

To test the Mathematical expression, the minimum test cases should be calculated according each parameter in it based on the method above. For example, to test the expression: X (operator) Y (operator) Z, in which the operator can be any mathematical operator, the minimum test cases will be based on the parameter: X, Y and Z, and make reasonable combinations to derive the test cases.

One of key tests about mathematical expression is the maximum and minimum value tests, for example, to test the expression: (X x Y) / Z about the maximum value, both parameters X and Y should use their maximum values with the minimum value for Z.

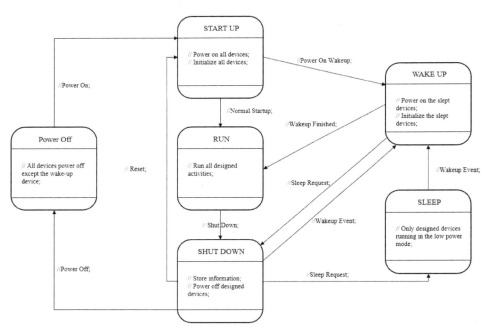

Figure 3.5-1 Mode Manager State Diagram

State Diagram

To test the control logics that are expressed by a state diagram, the minimum test cases should be:

- For each state, every expression should be tested according the expression and parameter testing methods above.

- For each transition, every transition path should be tested.
- To achieve the high testing quality, certain combinations of state and transition should be tested. For an example in Figure 3.5-1 Mode Manager State Diagram about the "Mode Manager State Diagram", from the "RUN" state to the states of "Power Off", "Sleep" and "Wake Up" via the state of "Shutdown", the transitions will travel two states, and the tests to those transitions will reveal the connections and the interactions during the transitions. So, the two states travel transitions should be tested to against the ASIL B or higher quality requirements.

3.5.1.2 *Test Case Content*

There are two goals for a test case:

- To fully reveal and verify the functionality of item under test.

A test case not only tests an item output, but also discovers what causes such outputs and in what environment and conditions, even the outputs may match the expectation, the result may not be correct if the environment and conditions are not right.

- To provide enough information about the test setup, especially the test case configuration, so that anyone else is able to have the same test result by executing the test case under the same configuration.

Such characteristic of test case is the test case consistency that is valuable to the regression test, and is useful to the failure analysis.

For example, if the development makes a software modification, and based on the impact analysis, the modification changes the function A, but not the function B, then the test cases for the function A and B from the previous development should provide such evidence that the function A is changed according to the modification and the function B is not changed if the test cases are configured based on the recorded configuration. In this way, the impact by the modification can be identified and isolated, which makes the regression test much efficient.

The approach to achieve the goals above is to design the test cases, which consists of:

- Test Target

This section should directly and clearly describe what the test targets are, so that by reading the section, the readers can quickly know what the test is about.

If the item under test is the ECU, then this section should describe which functionalities of the ECU will be tested and in what test environment.

If the item under test is the microcontroller, then this section should describe what functionalities represented by the chip port behaviors will be tested and in what test environment: on the target ECU board or the development board.

If the item under test is processor, then this section should describe what interactions and with what devices, such as RPU with APU, RPU with PMU, RPU with external ROM, will be tested.

If the item under test is a software function, then this section should describe which data, such as input data, output data, middle data, will be tested.

- Test Configuration

The test configuration should provide enough information about the test environment setup, so that anyone else is able to re-build the test environment and to repeat the test by following the information, which should consist of:

- o Equipment, such as devices, computer, assembly, vehicle
- o Software, such as Self developed test software, CANalyzer, Jira, dSpace, QAC
- o Data, such data file name, version and location, the detailed test data values are documented by the test data section below.

- Test Data and Output to be Recorded

The test data include not only the data inside of the item under test, but also the data to setup the test environment. The content of output to be recorded should describe what information will be recorded during the test, which not only covers the test output but also the test middle data that will help to analyze the test result.

- Test Execution

This section should have enough information about the test execution, such as, at what situation a break point is set, then what data are manipulated and continue to test, so that anyone else is able to repeat the test by following the steps:

- o It is efficient to the regression test.
- o It is the evidence to analyze if the test result is trustable.

- Test Criteria

This section should define the criteria to decide if the test case is passed, failed or Not Executed, the suggestions are to use the computer readable expressions to define the criteria, so that the results can be processed automatically using a software program, such as script code. Meanwhile, a test case is executed based on some other test case results, so the criteria should indicate that the test case should not be executed if based test case results do not meet the conditions, so that the current result should be Not Executed.

- Test Analysis and Result

This section should have the test result analysis based on the test records, and decide the test result: Passed or Failed according to the test criteria.

3.5.2　System Integration and Verification

The system integration and integration verification are the processes that:

- integrates all relevant system elements together to derive the required output result data values at the required time,
- verifies the derived data values and the timing against the design specifications.

In theory, the system integration should be smooth: all the system components are developed according to the requirement and design specifications, in which the functionalities and interfaces are specified accordingly, so they should be matched with each other. However, if the specifications themselves have issues, or the implementations have some deviations from the design, then the deviations and issues will be inherited into the result products, which will result in the two types of mismatches: Data Values and Data timing mismatches between the system elements, so, to ensure the data values and the data timing matches are the verification goals.

The system integration and integration verification must be done using the simulations because:

- the needed hardware and relevant software platforms are not ready.
- The execution environments that are simulated can be manipulated to verify the product in detail.

So, the simulation used to do the system integration and verification is the key to the successful integration tests, which needs to be as close as possible to the real environment and there should be all the necessary testing measurements to record, observe and manipulate the test data. And the system integration simulation is not only used in the system integration test, but also used in the verification system operation concept which is developed at the very beginning of the development, which is very valuable, and enable the development can be done in parallel during the system architecture design, that is the reason why that the system verifications don't have to be done at the late phase of the development.

Among the two verification goals: data value and timing, the result data timing is difficult to be verified because the integration environment is the simulation, not the real environment, which will generate the issues that the simulated environment is not close enough to the real one, which results in that the result timing is not accurate. And the simulation is impossible to be exactly same as the real execution environment because there are the testing measurements in both the component under test and the integration environment to observe the internal information to make the judgments about the results.

The data value verification can be done by following the data flows in the system operation concept, in which, all the data and the relationships to derive the final result data are in the concept, so, the integration and the integration verification just need to follow the concept to integrate those elements, and verify the results against the specifications.

The system integration and integration verification can be divided into following phases:

- First Phase: Simulation verification, which focuses on:
  - o The operation concept, i.e., to verify the result data derivations based on the system operation concept.
  - o The data types and interfaces that are used to derive the result data.

3.5.2 System Integration and Verification

- Second Phase: Integration and Verification based on Hardware Platform, which focuses on each hardware platforms, such as APU, RPU processors, all the software components in the processor will be integrated together and will be verified against their interactions, in which, the following aspects will be verified:
  - o The processor start and shutdown,
  - o The partitions in the processor start and shutdown,
  - o The communications between the software functions,
  - o The processor input and output signal handlings,
  - o The resources in the processor usage, RAM, ROM,
  - o The processor execution, such as scheduler, watchdog, interrupts
- Third Phase: Microcontroller, in which, all software components for every processor in the microcontroller will be integrated and verified:
  - o The microcontroller start and shutdown,
  - o The devices in the microcontroller start and shutdown,
  - o The communications between the devices,
  - o The microcontroller input and output signal handlings,
  - o The resources usage, such as external RAM, external ROM, external watchdog, external clock signal,
  - o The microcontroller execution, such as the APU, RPU, SHE, those application software platforms' execution should cooperate with microcontroller platform management unit (PMU), the memory management unit (MMU) and the peripheral management unit (PMU)
  .
- Last Phase: ECU, in which, all the devices in the ECU, such as the power supply unit, communication interface devices, the ECU on board temperature measurement circuits, the vehicle power supply measurement circuits and other need devices and circuits, will be integrated and verified:
  - o The ECU power on and off including the ECU sleep and wakeup,
  - o The communications with the vehicle including the network management,
  - o The ECU input and output signal handlings,
  - o The ECU environment interaction, such as temperature measurement, vehicle power supply measurement,
  - o The ECU software update using bootloader,
  - o The ECU testing and maintenance using JTAG or other interfaces.

The following is an example of first phase to verify the system by following the system operation concept, which focuses on if the data values match each other between the system elements, data flow and operation logics in the system operation concept:

Object_Detected (Object_Relative_Position, Object_Relative_Velocity) = f_detection
(

  Pixel_Mapping,
Object_Classification_Array,
Frame_Input_Array,
Object_Detect_Time_Constant,
CAN_Msg_Veh_Speed);

Among the parameters, each of them needs to be integrated into the formula respectively:
- Pixel_Mapping,
- Object_Classification_Array,
- Frame_Input_Array,
- Object_Detect_Time_Constant,
- CAN_Msg_Veh_Speed.

First step: Integrating the Pixel_Mapping, by simulating the other three parameters, then input the integrated target data by setting a few design values of Pixel_Mapping according to the minimum test cases mechanism to check if the result data: Object_Relative_Position and Object_Relative_Velocity meet the calculation logic.

Second step: Integrating each of other three parameter individually using the same procedure above by designing the minimum test cases.

Third step: continuously changing the value of the Pixel_Mapping by simulation, then to check if the result data: Object_Relative_Position and Object_Relative_Velocity meet the calculation logic.

Forth step: continuously changing the values for each of other three parameters individually, then to check the output result.

The integration method above is suitable to check if the data values meet the specifications, however, it is not suitable to check if the data timing meets the specification because the four elements in the example above may not be in the same partition or same processor, so, the communication timing will be critical for the integration if the elements are allocated in the different partitions or processors.

Then, next is the second phase: Integration and Verification based on Hardware Platform, which is to verify the execution events and the operation sequence that derive the result data in a specific processor, in which, the system elements are classified as below integration levels:
- Software Function in a partition
- Partition
- Processor (or Core)

So, this part of system level integration should start from the software function level to integrate all the software functions in a partition, then integrate all partitions in a processor. For example, to integrate the contents in the process of AUP0 that is described in the section of "Safety at the partition level" of 3.4.6.3 BSD Safety listed as below:
APU0:
- Only one partitions, which is ASIL B;
- Object_Detected (Object_ID, Object_Label) = f_classification (
    Pixel_Mapping,
    Object_Classification_Array,
    Frame_Input_Array,
    Object_Detect_Time_Constant);
- Object_Detected (Object_Relative_Position, Object_Relative_Velocity)
    = f_detection (
    Pixel_Mapping,
    Object_Classification_Array,
    Frame_Input_Array,
    Object_Detect_Time_Constant,

CAN_Msg_Veh_Speed);
- Autosar: OS, EcuM, BswM;

In which, the contents of Object_Detected, Object_Relative_Position and Object_Relative_Velocity that are described in the first phase are calculated here together with the contents of Object_Detected, Object_ID and Object_Label. And to manage those calculation, executions and the resources, such RAM, ROM and devices, the AUTOSAR components: OS, EcuM and BswM are implemented in the processor, as well. All those contents are allocated in one partition that is the ASIL B.

In such integration and integration verification based on a processor, the Hardware in Loop (HIP) and the Software in Loop are helpful. For example, in the calculation of Object_Detected (Object_Relative_Position, Object_Relative_Velocity, the signal: CAN_Msg_Veh_Speed is needed, if the CAN interface is ready at the time, then signal can be integrated using the target hardware device, then that is the HIL; and if the CAN signal hardware is not ready, but the CAN signal software is ready and can run on alternative hardware like CANalyzer, then the signal can be integrated this way that is the SIL.

The similar HIL or SIL approach can be used to integrate the data of Pixel_Mapping that is from the GPU if the signal is not ready at the time.

Once all the software for all the processors in a microcontroller are ready, then the next step of integration and integration verification is to integrate all those software in the target microcontroller, which is usually done on the device called the development board, which has the target microcontroller with the necessary device and circuit to run the microcontroller, but it does not have other target devices, such as the power supply unit, external RM and ROM. In this way, the contents in the microcontroller can be fully verified.

The last step is to run all target software and hardware devices on the target ECU module, in which, all the focus points mentioned above need to be verified, and pay the special attention to the contents described in the section of State Control and Execution Monitor in 3.4.6.3 BSD Safety.

3.5.3 System Black Box Verification

The system black box verification that is sometimes called the system test is only about two types of tests:

- Time persistence test that focuses on:
  - Real operation environment
  - Power on and power off cycle including the sleep and wakeup cycle
  - System in fault status and recovery from fault status
  - Lifetime running test
  - Worst case threshold tests to make the workloads to reach the designed maximum values, such as the input signals from vehicle CAN bus come at the maximum rate, so that the system capacity can be verified; or to apply all possible CAN input signals including both valid and invalid signals on the CAN bus interface, so that the interface resistance ability can be tested, which is call "fuzz" test.

- Environment persistence test, which are against:
  - Electronic test:
    - Voltage test including the high and low voltages in the working range, and the possible voltage values out of the working range.
    - Electromagnetic interference.
    - Wire and wireless signal inference including both valid and invalid signals, which covers the fuzz test.
  - Temperature test:
    - High temperature in the working range.
    - Low temperature in the working range.
    - Thermal Shock test that the temperature out of working range.
    - Thermal humidity test.
    - Solar radiation soak test.
  - Liquid test:
    - Water or steam intrusion, saltwater immersion, salt fog resistance, mud resistance, chemicals resistance that may existed in the vehicle,
  - Physical test:
    - Vibration, mechanical shock, handling drop, gravel bombardment, sand resistance.

Note: the vehicle tests are out of ECU suppliers' scope, rather it is the OEMs' responsibility. The ECU development should be based on the OEMs' specifications which need to be precisely specified from the EU black box point of view.

In the system black box verification, the software should be the real production software, not the development software with the diagnostic functions, because the time persistence test should verify if the output data timing meets the requirements, so any non-production software function may impact the process timing, which may in turn impact the output data timing.

If the configurations can be modified by updating the configuration file using the bootloader, then all the possible configurations should be tested by changing the data files.

If the data or parameters can be modified by using the diagnostic services, then all the possible data or parameters and their feasible combinations should be tested.

The test cases can be easily designed by combining the input signals using both valid and invalid values, in which, the normally required fuzz test will be covered. For which, the information that is managed using the requirement management tool, such as in the Table 3.3-2 BSD Input & Output Signal, can be used conveniently to test the system. From which, the tests for the combinations of input signals will duplicate some test cases that are covered by the integration tests, which is the necessary duplicate because in the black box test, there is not any testing software, so that the timing tests will be accurate.

Check List 10 - System Verification

- Preparation
  - Are the requirement and design specifications clearly defined and fully understood?
  - Are the system verification scope and approach defined?
  - Are the operation logics of the device under test (DUT) clearly defined and expressed?
  - Are the test cases designed based on the operation logics above?
  - Are the regression test cases designed based on the regression strategy to focus on the impacted functions?
  - Do the test cases granularity meet the device under test (DUT) critical characteristics including importance, safety and cybersecurity?
  - Do the test cases have enough detailed information including the configuration, stimulus and middle data to fully test the device under test (DUT)?
  - Do the test cases have enough detailed information so that they can be reliably repeated by somebody else?
  - Does the test setup have the ability to trigger the stimulus and record the outcome information?
- Execution
  - Are all the actions in the test controlled by a device, not by manual (so that the actions can be repeated precisely)?
  - Are all the output data tested based on the operation concept?
  - Are all the states tested (startup, shutdown, initial, normal, fault, recovery, sleep, wakeup)?
  - Are all the software components tested (Feature Function, Application Mode Manager, Serial Signal Manager, Diagnostic Service, Cybersecurity Function, AUTOSAR)?
  - Are all hardware devices tested (partition, processor, microcontroller, power supply, communication interfaces, external RAM and ROM, I/O circuits)?
  - Are all features tested (ECU feature, functional safety, cybersecurity, bootloader, maintenance function)?
  - Are all the configurations, parameters and all the possible situations tested (temperature, voltage, physical environment, signal interference)?
- Result Analysis
  - For the test case that test the output data: if the output data match the expected results, do the middle data match the expected values, as well? And does the output data timing meet the expected time?
  - For the test case that test the states: if the state transitions match the expected results, do the transition conditions match the expected conditions, as well? And does the state transition timing meet the expected time?
  - Do the test results match the impact analysis (both changed functions and non-changed function match the expectations)?

3.5 System Verification

Appendix A: Check List

Appendix B: Figure Index

Appendix C: Table Index

www.ingramcontent.com/pod-product-compliance
Lightning Source LLC
LaVergne TN
LVHW081521050326
832903LV00025B/1567